Algebra Tools

Mathematics in Context

Algebra

HOLT, RINEHART AND WINSTON

Mathematics in Context is a comprehensive curriculum for the middle grades.
It was developed in 1991 through 1997 in collaboration with the Wisconsin Center
for Education Research, School of Education, University of Wisconsin-Madison and
the Freudenthal Institute at the University of Utrecht, The Netherlands, with the
support of the National Science Foundation Grant No. 9054928.

This unit is a new unit prepared as a part of the revision of the curriculum carried
out in 2003 through 2005, with the support of the National Science Foundation
Grant No. ESI 0137414.

National Science Foundation
Opinions expressed are those of the authors
and not necessarily those of the Foundation.

Kindt, M, Dekker, T., Burrill, G., and Romberg T. A. (2006). *Algebra tools*.
In Wisconsin Center for Education Research & Freudenthal Institute (Eds.),
Mathematics in Context. Chicago: Encyclopædia Britannica, Inc.

ISBN 0-03-040387-1

3 4 5 6   073   09 08 07 06 05

The *Mathematics in Context* Development Team

Development 2003–2005

Algebra Tools was developed by Martin Kindt and Truus Dekker. It was adapted for use in American schools by Gail Burrill and Thomas A. Romberg.

Wisconsin Center for Education

Research Staff

Thomas A. Romberg
Director

David C. Webb
Coordinator

Gail Burrill
Editorial Coordinator

Margaret A. Pligge
Editorial Coordinator

Project Staff

Sarah Ailts
Beth R. Cole
Erin Hazlett
Teri Hedges
Karen Hoiberg
Carrie Johnson
Jean Krusi
Elaine McGrath

Margaret R. Meyer
Anne Park
Bryna Rappaport
Kathleen A. Steele
Ana C. Stephens
Candace Ulmer
Jill Vettrus

Freudenthal Institute Staff

Jan de Lange
Director

Truus Dekker
Coordinator

Mieke Abels
Content Coordinator

Monica Wijers
Content Coordinator

Arthur Bakker
Peter Boon
Els Feijs
Dédé de Haan
Martin Kindt

Nathalie Kuijpers
Huub Nilwik
Sonia Palha
Nanda Querelle
Martin van Reeuwijk

Cover photo credits: (left) © Getty Images; (middle, right) © Corbis

Illustrations
2, 3, 5, 45, 46 Christine McCabe/© Encyclopædia Britannica, Inc.

◆ Contents

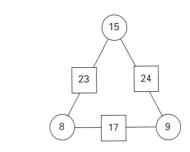

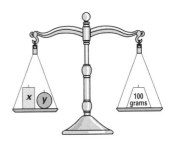

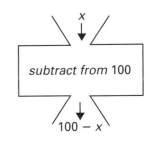

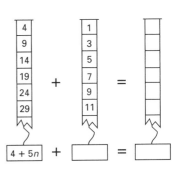

◆ Contents

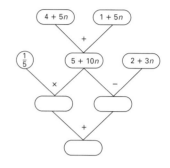

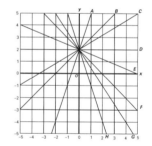

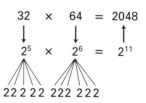

◆ Contents

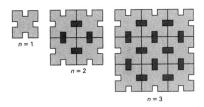

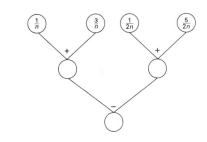

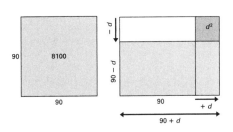

Peanut Butter and Milk

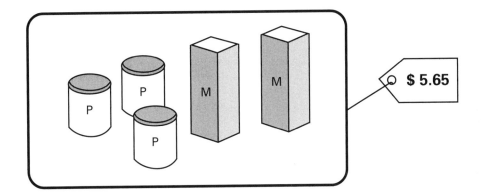

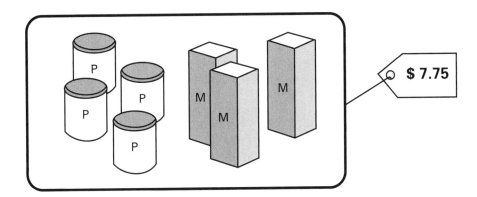

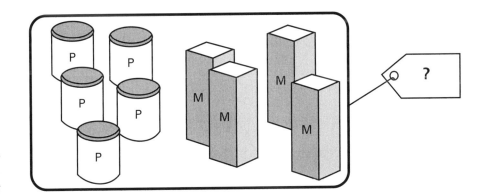

- Explanation.

Comparing Prices (2)

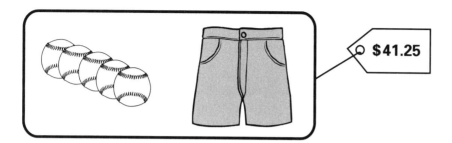

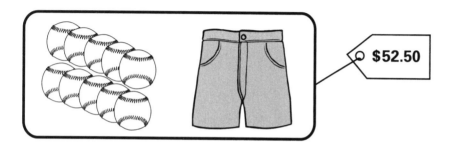

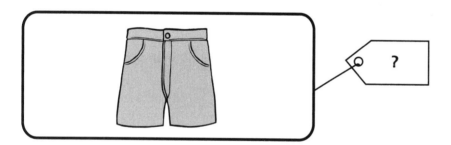

- Show your work.

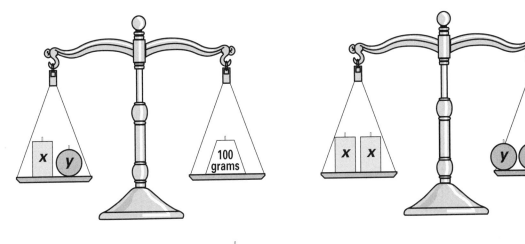

x _____ grams

y _____ grams

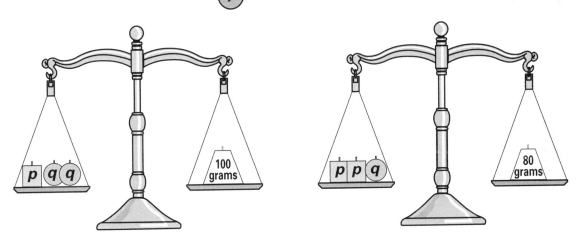

p _____ grams

q _____ grams

Comparing Lengths

There are four pieces of rope. Three pieces have the same length; the fourth one is longer than the other three.

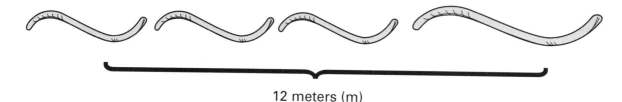

12 meters (m)

All the pieces together are 12 m.

One short and one long piece together are 7 m.

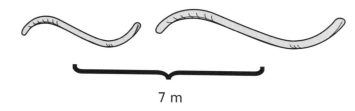

7 m

- What is the length of one short piece of rope? One long piece? Show your work.

Another rope, with a length of 18 m, is cut in 5 pieces.

Two pieces are equal in length. The other three pieces are equal in length, but they are shorter than the first two.

One short and one long piece are 8 m together.

- What is the length of one short piece of rope? One long piece? Show your work.

together 54

together 21

together 51

How Long?

Three pieces of rope

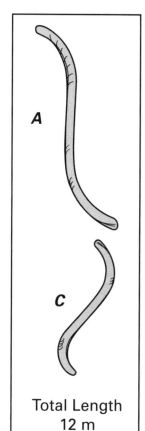

Total Length
12 m

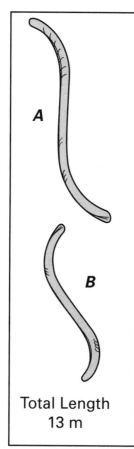

Total Length
13 m

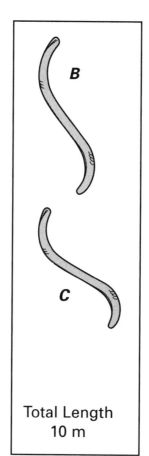

Total Length
10 m

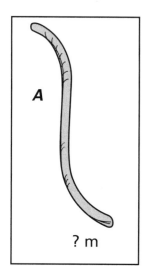

? m

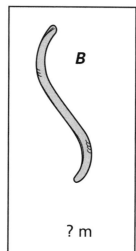

? m

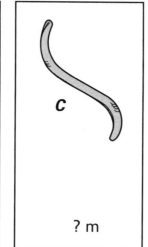

? m

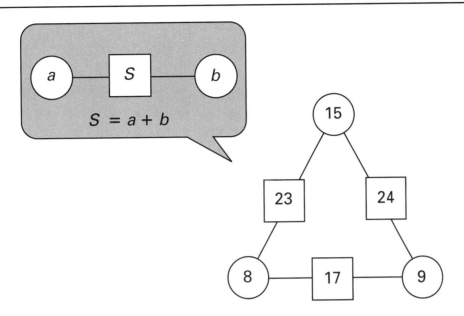

$S = a + b$

• Fill in the missing numbers.

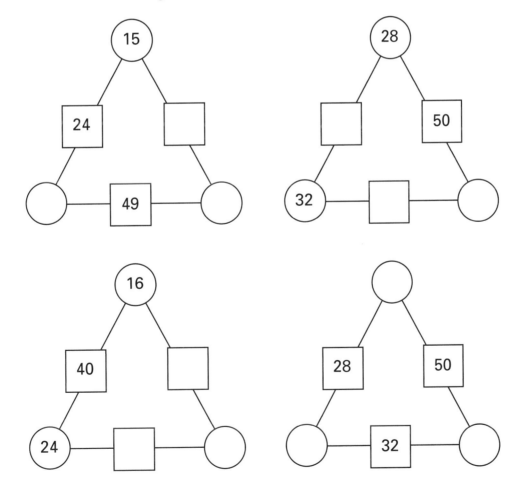

Find the Missing Numbers

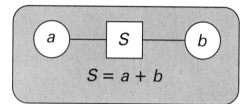

$$S = a + b$$

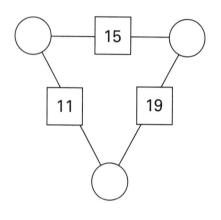

• There is only one solution. Which one?

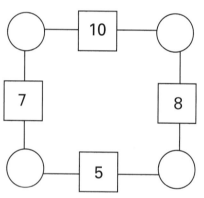

• There are a lot of solutions!

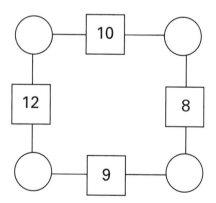

• This doesn't work! Explain why not.

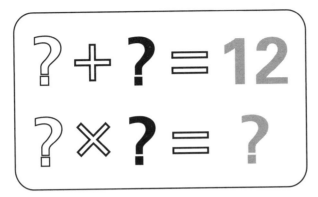

There are two positive integers. If you add both numbers, the result will be 12.

- What are the results if you multiply the integers?
  Show all the results you found.

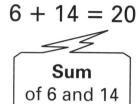

Sum and Product (2)

$$6 + 14 = 20 \qquad 6 \times 14 = 84$$

Sum
of 6 and 14

Product
of 6 and 14

Two positive integers have a sum of 20.

If the numbers were 6 and 14, their product would be 84.

But there are other pairs of positive whole numbers whose sum is equal to 20 with products that are not 84.

Below you see a chart with numbers from 1 to 100.

- Color all the cells that show a product for two numbers that have a sum of 20.

1	2	3	4	5	6	7	8	9	10
11	12	13	14	15	16	17	18	19	20
21	22	23	24	25	26	27	28	29	30
31	32	33	34	35	36	37	38	39	40
41	42	43	44	45	46	47	48	49	50
51	52	53	54	55	56	57	58	59	60
61	62	63	64	65	66	67	68	69	70
71	72	73	74	75	76	77	78	79	80
81	82	83	84	85	86	87	88	89	90
91	92	93	94	95	96	97	98	99	100

- Write the results in a sequence from large to small.

- Do you see any pattern in this sequence? Describe the pattern you see.

Mathematics in Context

Two positive whole numbers have a product equal to 24.

- What can their sum be?

Three positive whole numbers have a sum equal to 10.

- What can their product be?

Sum and Product (4)

A and *B* represent positive whole numbers.

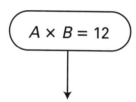

$A + B = 12$

$A \times B =$ _____ or _____ or _____ or _____ or _____ or _____

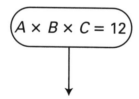

$A \times B = 12$

$A + B =$ _____ or _____ or _____

$A \times B \times C = 12$

$A + B + C =$ _____ or _____ or _____ or _____

> **Number Game**
>
> Think of a number less than 10. This is your starting number.
> Add 10 to your starting number.
> Also subtract your starting number from 10.
> Add both results. Which number do you get?
> Repeat this a few times with another starting number.

- What happens? How can you explain this?

Change the number 10 to 25 and play the same number game.
- What will the result be?

Sum and Difference (2)

The sum of two numbers is equal to 20.

If both numbers are equal, each of them is 10.

If the numbers are unequal, the bigger one is greater than 10.

So this number is **10 + any number** or **10 + a** for short.

- How can you write the smaller number for short?

Look at the number line.

The sum of two *unequal* numbers is 60.

The bigger number is indicated on the number line.

- Indicate the smaller number on the number line.

  Be as accurate as you can.

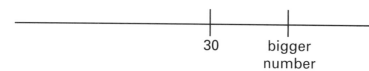

30   bigger
        number

You know that the sum of two numbers is 100 and their difference is 12.

- Find these numbers.

You know that the sum of two numbers is 80 and their difference is 15.

- What are these numbers?

Isabelle earned $100. She wants to buy a pair of running shoes. They normally cost $70. So she expects $30 will be left.

She is lucky! When she enters the shop, she discovers that the pair she wants is reduced by $8.

- How many dollars does she have left?

There are two ways to calculate this.

    a. $100 - (70 - 8)$

    b. $(100 - 70) + 8$

- Did you use method **a** or method **b**?
- Without looking at the result of the calculation, explain why the method you did not use would work just as well.

Suppose Isabelle already knew that the price of the shoes was reduced, but she did not know how much. So she knew that more than $30 should be left!

- Use the story to explain $100 - (70 - a) = 30 + a$.

A more general equality is $100 - (p - a) = 100 - p + a$.

- Explain this equality.

    (You may suppose $p < 100$ and $a < p$.)

A frequently occurring error is $100 - (p - a) = 100 - p - a$.

- Invent a short story in which someone has $100 - p - a$ left from 100 dollars instead of $100 - p$.
- Fill in the right expression: $100 - (\underline{\hspace{2cm}}) = 100 - p - a$.

Different Differences (2)

If you subtract less, there will be more left.

Example:

$$50 - 20 = 30$$
$$50 - (20 - x) = 30 + x$$

If you subtract more, there will be less left.

- Invent an example.

- Here are four conclusions. The first one is complete. Explain that one.
- Complete the other three.

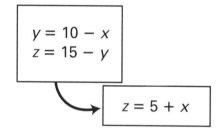

$y = 10 - x$
$z = 15 - y$

$z = 5 + x$

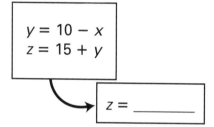

$y = 10 - x$
$z = 15 + y$

$z = $ _____

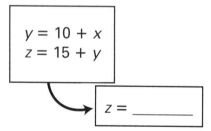

$y = 10 + x$
$z = 15 + y$

$z = $ _____

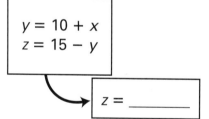

$y = 10 + x$
$z = 15 - y$

$z = $ _____

- Invent some other conclusions in the same style.
  Use symbols other than x, y, and z.

Mathematics in Context

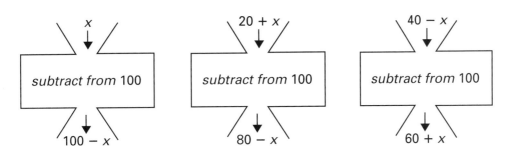

- The sum of INPUT and OUTPUT is 100 in each case. Check this out.
- Complete the chain.

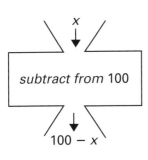

The chain on the left corresponds to the following chain of differences.

$$30 - [80 - (150 - (100 - x))] = \underline{\qquad}$$

- This expression is equivalent to an expression with only one term. What is this expression?

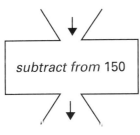

- Make chains corresponding to:

$$10 - (9 - (8 - y))$$
$$16 - [9 - (4 - (1 - a))]$$
$$32 - [16 - [8 - (4 \quad (2 - k))]]$$

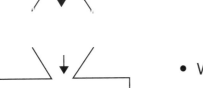

- What are the three resulting expressions?

- Invent a "chain exercise" yourself. Provide an answer to it.

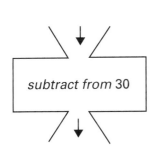

Find the Value of *x* (1)

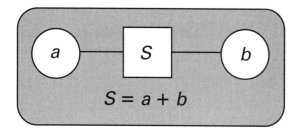

$S = a + b$

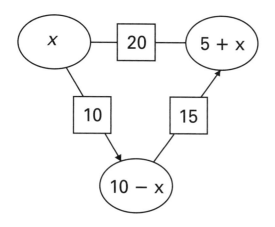

$x + 5 + x = 20$

$x =$ _____

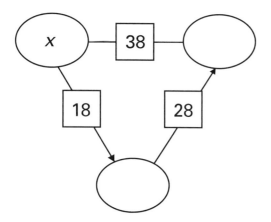

_____

$x =$ _____

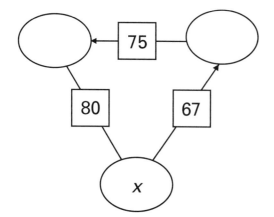

_____

$x =$ _____

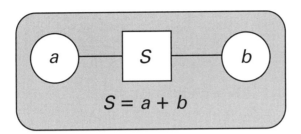

$$S = a + b$$

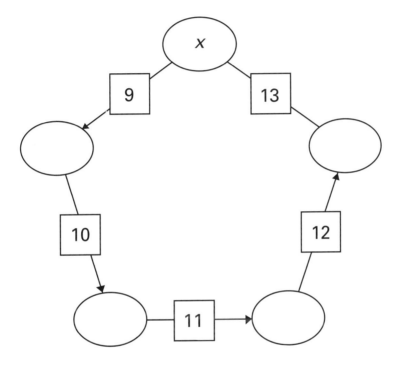

_____

x = _____

Difference in Temperature

All temperatures are in degrees Celsius.

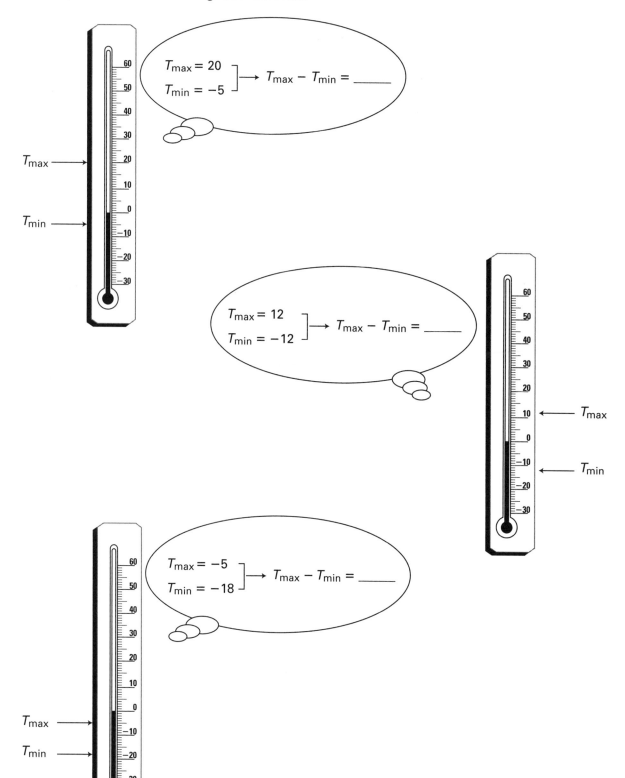

$T_{max} = 20$
$T_{min} = -5$ $\rightarrow$ $T_{max} - T_{min} =$ _____

$T_{max} = 12$
$T_{min} = -12$ $\rightarrow$ $T_{max} - T_{min} =$ _____

$T_{max} = -5$
$T_{min} = -18$ $\rightarrow$ $T_{max} - T_{min} =$ _____

Positive and Negative (1)

• Fill in the missing numbers.

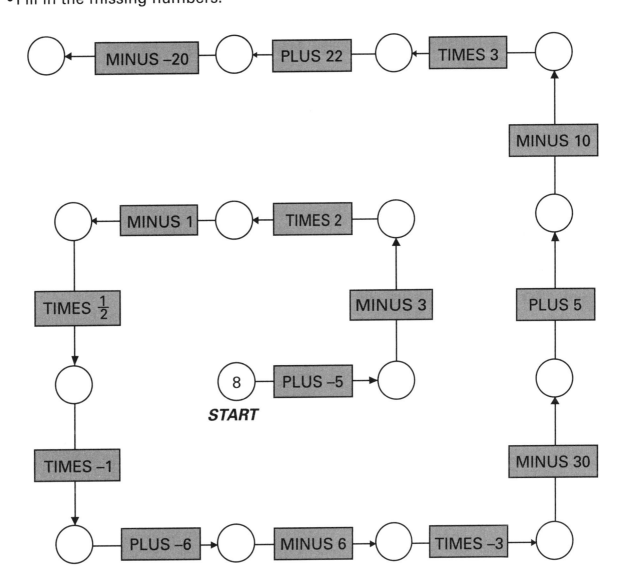

Positive and Negative (2)

• Fill in the missing numbers.

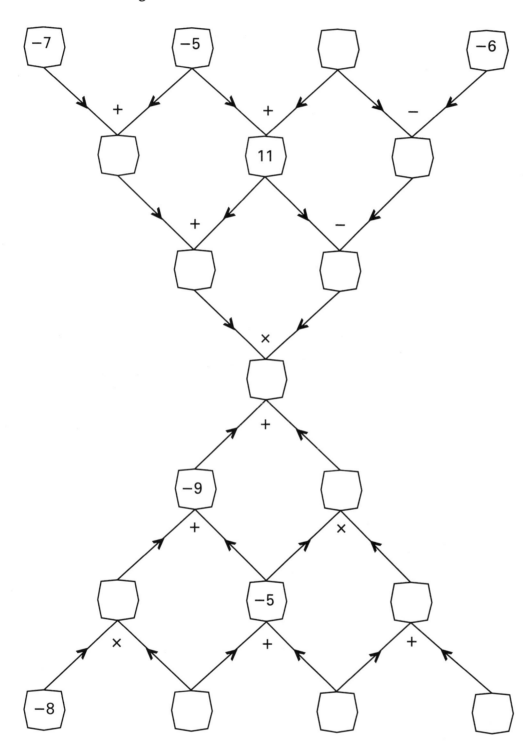

Mathematics in Context

$(25 - 28) - (32 - 40) =$ _____

$25 - (28 - 32) - 40 =$ _____

$25 - (28 - (32 - 40)) =$ _____

$25 - (28 - 32 - 40) =$ _____

$25 - 28 - (32 - 40) =$ _____

$25 - 28 - 32 - 40 =$ _____

Positive and Negative (4)

A and *B* are integers.

$$A + B = 5$$

$$-6 < A < 6$$ ← This means that *A* has a value between −6 and 6.

- What can the value of *B* be? Of *A* − *B*? Of *B* − *A*?
- Fill in the chart.

A	B	A − B	B − A
−5			
−4			
−3			
−2			
−1			
0			
1			
2			
3			
4			
5			

Look at the point (3, 2).

The sum of the *x*- and *y*- coordinates is 5, as is indicated by the flag.

A diagonal line is drawn through (3, 2).

- Check that the *x*-and *y*-coordinates of every grid point on this diagonal line have the same sum.

Between the grid points there are points that have fractions as coordinates.

- Indicate three such points on the diagonal line and check whether the sum of their coordinates is also 5.

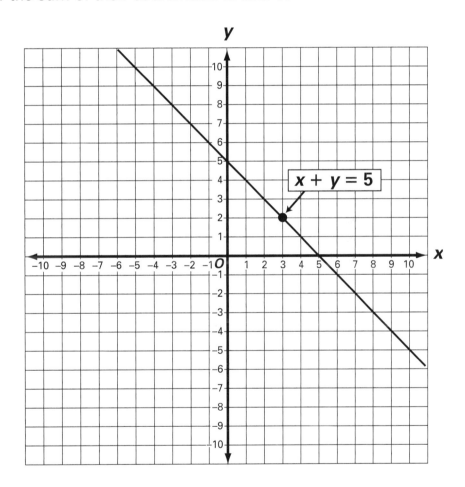

$x + y = 5$

- Draw the diagonal line going through the points with coordinate sum −5. Label the line with **x + y = −5**.
- Draw a diagonal line parallel to the two diagonals already there so that it lies exactly in the middle between them. What label should you give this line?

Diagonals of the Grid (2)

Look at the point (7, 2).

The difference of the *x*- and *y*-coordinates is 5.

- Draw the diagonal line through (7,2) so that the *x*- and *y*-coordinates of every grid point out the diagonal line have the same difference and label it $x - y = 5$.

- Do the same, but now for the diagonal $y - x = 5$.

- Draw two other diagonals parallel to both lines and label each of them.

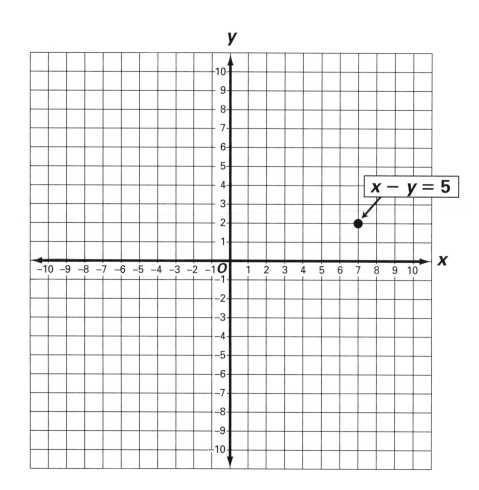

$$x - y = 5$$

- Draw the diagonals of the grid with labels $x + y = 7$ and $x - y = 3$.
  What are the coordinates of their point of intersection?

- Draw two diagonals going through the point $(-4, 2)$.
  How can you label these lines?

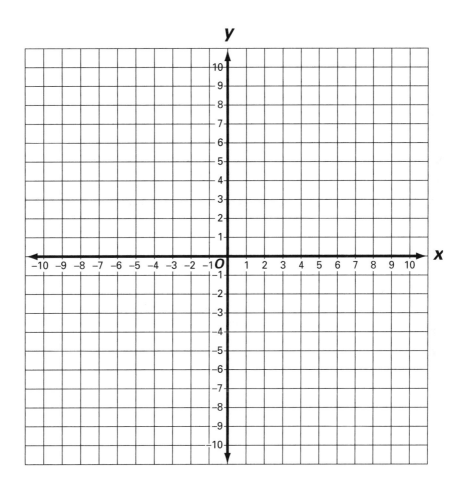

Name _____ Date_____ Class_____

Multiplying on the Grid

Look at the arrow (**A**) in the grid. This figure has 7 vertices.
- Give the coordinates of these vertices.

If you multiply the *x*-coordinates of the vertices of **A** by –2,
you get a new arrow.
- Draw that new arrow.

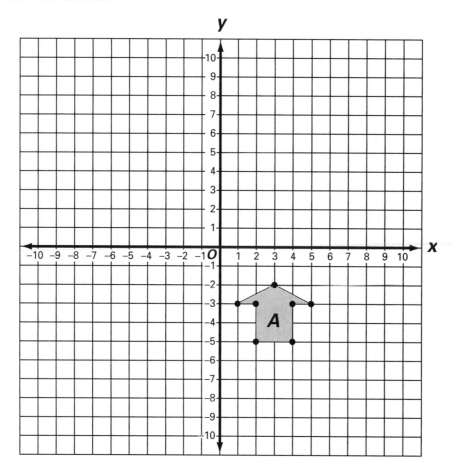

- Now multiply only the *y*-coordinates of **A** by –2 and draw
  the new arrow you get.

- Multiply both coordinates of **A** by –2 and draw the new
  arrow.

Mathematics in Context

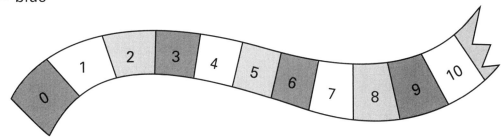

■ = red

□ = blue

The number strip has a repeating pattern of
red-white-blue-red-white-blue, etc.

• What color is the cell of the number of your age?

• What color is the cell of number 67? Of 667?

• Explain why you can be sure that the color of the
number 111 is red.

• What will the color of the number 1,111 be?

The numbers in the red cells can be described by
a formula:

red number = 3 × n

• Explain this formula.

• Give a similar formula for the white numbers

• Give a similar formula for the blue numbers.

Colored Numbers (2)

This number strip has five colors.

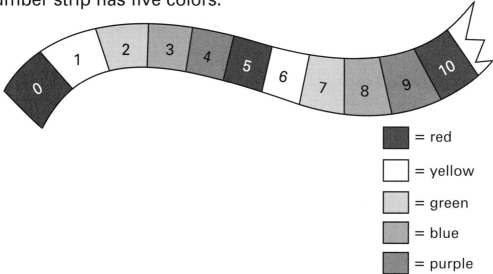

☐ = red
☐ = yellow
☐ = green
☐ = blue
☐ = purple

- Write all the red numbers between 21 and 38.

- Write the first even yellow number after 100.

- How do you know that the number 99,999,999 is purple?

- Give a formula for the green numbers.

If you add a green and a blue number, the result will be red.
- Explain why.

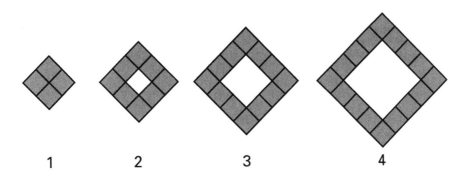

1   2   3   4

- How many gray squares will the pattern with number 5 have? The pattern with number 25?

- Give a formula for the number of gray squares (**R**) expressed in the number of the pattern (**n**).

$$R = \underline{\hspace{2cm}}$$

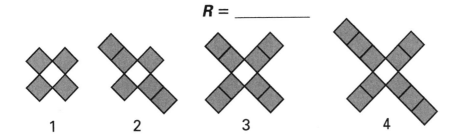

1   2   3   4

- Give a formula for the number of gray squares (**R**) expressed in the number of the pattern (**n**).

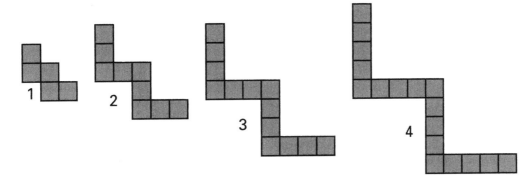

- Give a formula for the number of gray squares (**R**) expressed in the number of the pattern (**n**).

Patterns of Stars

1   2   3   4

- Give a formula for the number of stars (**S**) expressed in the number of the pattern (**n**).

S = _____

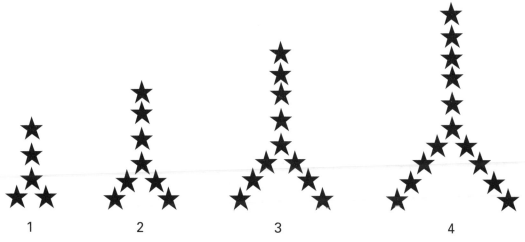

1   2   3   4

- Give a formula for the number of stars (**S**) expressed in the number of the pattern (**n**).

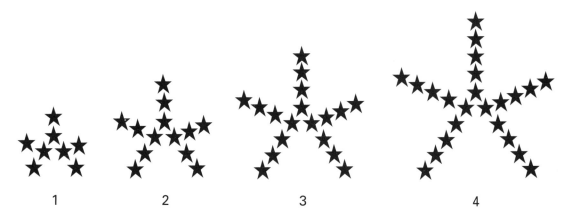

1   2   3   4

- Give a formula for the number of stars (**S**) expressed in the number of the pattern (**n**).

Arrow Strings (4)

• Fill in the missing expressions.

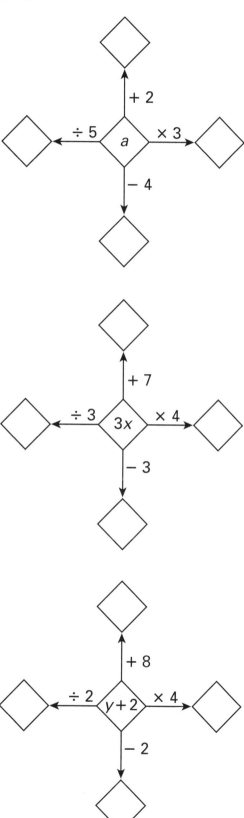

Mathematics in Context

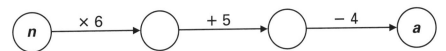

• Give a formula that corresponds to the first arrow string.

$$a = \underline{\hspace{2cm}}$$

You can change the order of the three "operators" ×6, +5, −4.
This leads to five new arrow strings.

• Give a formula to each of these five arrow strings.

$b = \underline{\hspace{6cm}}$

$c = \underline{\hspace{6cm}}$

$d = \underline{\hspace{6cm}}$

$e = \underline{\hspace{6cm}}$

$f = \underline{\hspace{6cm}}$

Arrow Strings (2)

- Fill in the missing numbers.

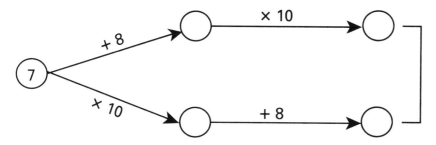

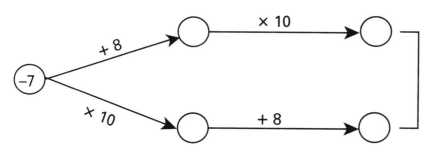

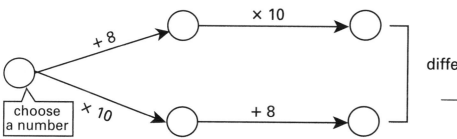

- Fill in the missing expressions.

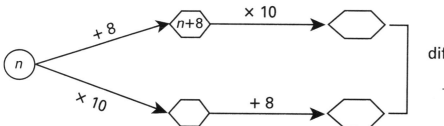

- Fill in the missing numbers.

i

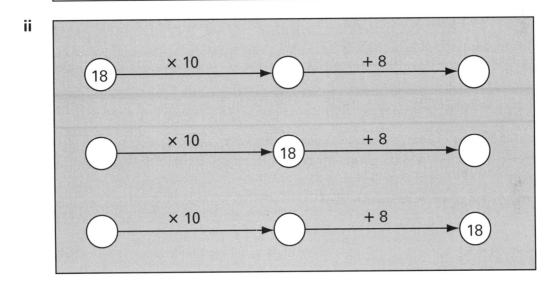

ii

Look at these two formulas.

A = 10 × n + 8
B = 10 × (n + 8)

- Which of the two formulas corresponds to the arrow string?

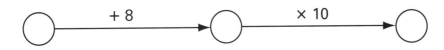

• Fill in the missing expressions.

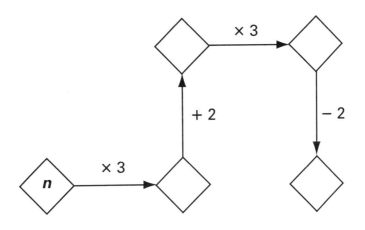

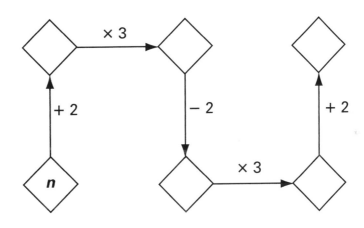

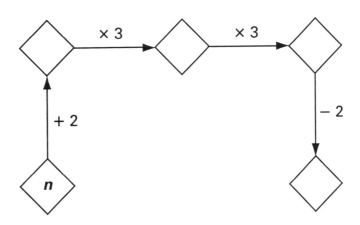

Operating with Number Strips (1)

• Fill in the missing numbers and expressions.

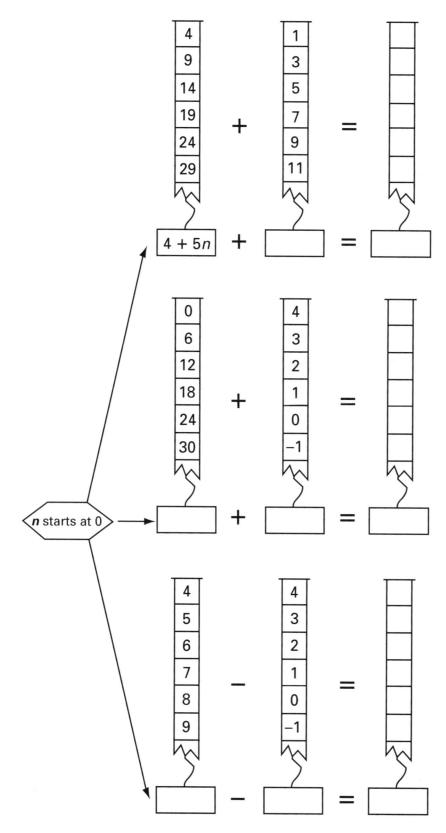

Operating with Number Strips (2)

- Fill in the missing numbers and expressions. In each case *n* starts at 0.

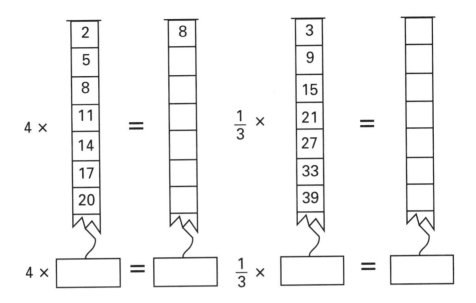

$4 \times$ [____] = [____] $\frac{1}{3} \times$ [____] = [____]

$8 \times$ [____] + $3 \times$ [____] = [____]

Strips and Charts (1)

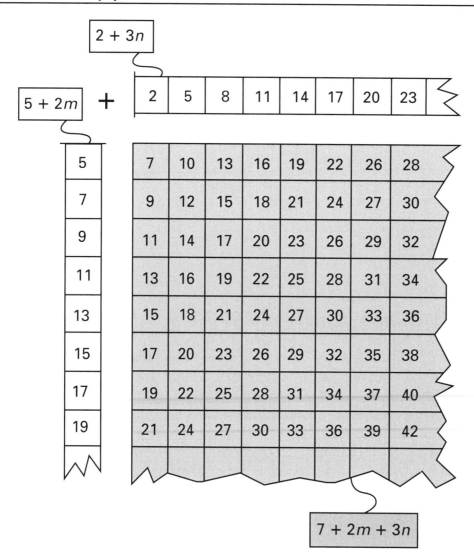

$2 + 3n$

$5 + 2m$ **+**

| | 2 | 5 | 8 | 11 | 14 | 17 | 20 | 23 |

5	7	10	13	16	19	22	26	28
7	9	12	15	18	21	24	27	30
9	11	14	17	20	23	26	29	32
11	13	16	19	22	25	28	31	34
13	15	18	21	24	27	30	33	36
15	17	20	23	26	29	32	35	38
17	19	22	25	28	31	34	37	40
19	21	24	27	30	33	36	39	42

$7 + 2m + 3n$

- Find the number in the chart that corresponds to $m = 3$ and $n = 2$.

- Find the number that corresponds to $m = 3$ and $n = 5$.

- Which horizontal row in the chart does $m = 4$ corresponds to?

- Which vertical column does $n = 5$ correspond to?

- Which numbers in the chart correspond to $m = n$?

- Make a number strip for the numbers that correspond to $m = n$ with a corresponding expression.

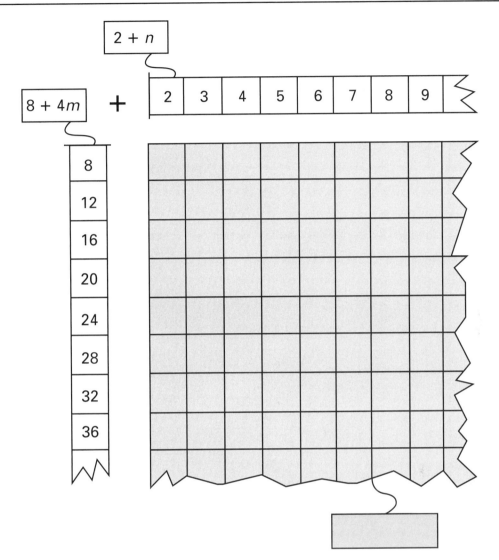

- Fill in the chart. What expression fits the chart?

- What strip fits **m = 3**? What is the corresponding expression?

- Answer the same questions for **n = 0**.

- For **m = n**.

- For **m = n + 1**.

Strips and Charts (3)

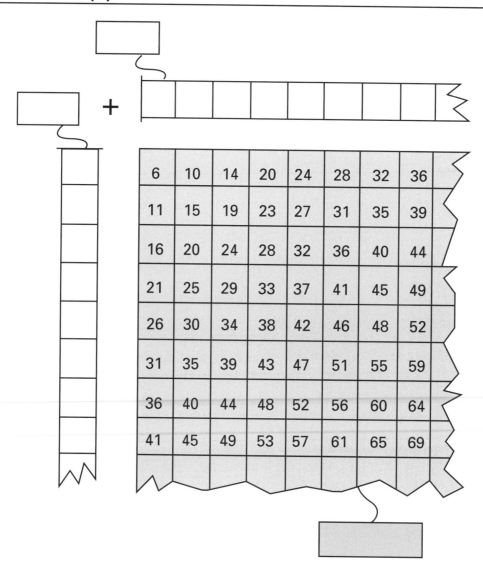

6	10	14	20	24	28	32	36
11	15	19	23	27	31	35	39
16	20	24	28	32	36	40	44
21	25	29	33	37	41	45	49
26	30	34	38	42	46	48	52
31	35	39	43	47	51	55	59
36	40	44	48	52	56	60	64
41	45	49	53	57	61	65	69

• What expression fits the chart?

• What strips can be used to make the chart?

Mathematics in Context

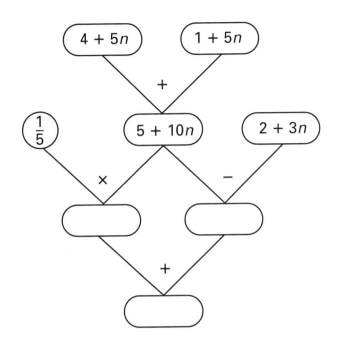

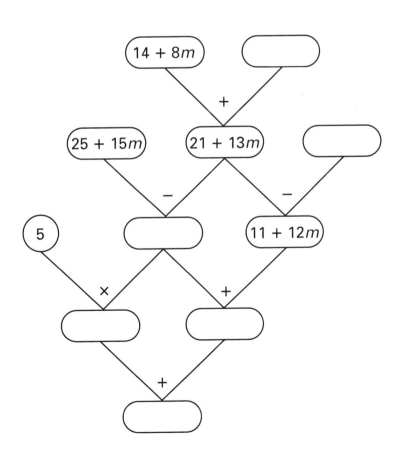

Operating with Expressions (2)

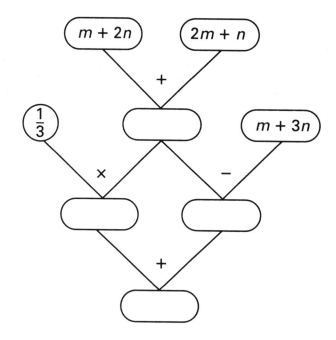

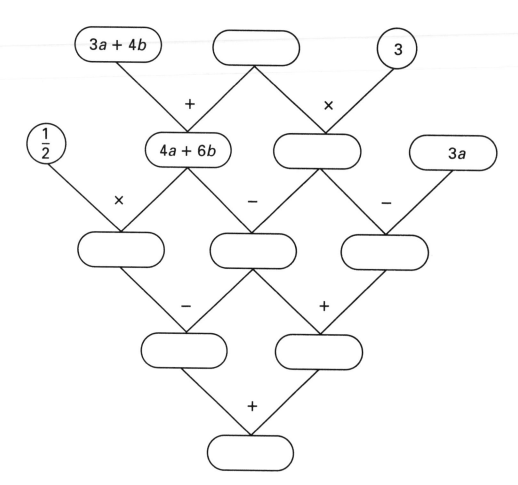

Name _____ Date_____ Class_____

Algebra in Balance (1)

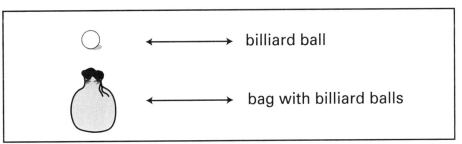

○ ←——————→ billiard ball

🛍 ←——————→ bag with billiard balls

Each bag contains the same number of billiard balls.

• Calculate this number. Ignore the weight of the bags.

© Encyclopædia Britannica, Inc. This page may be reproduced for classroom use.

Mathematics in Context

Algebra in Balance (2)

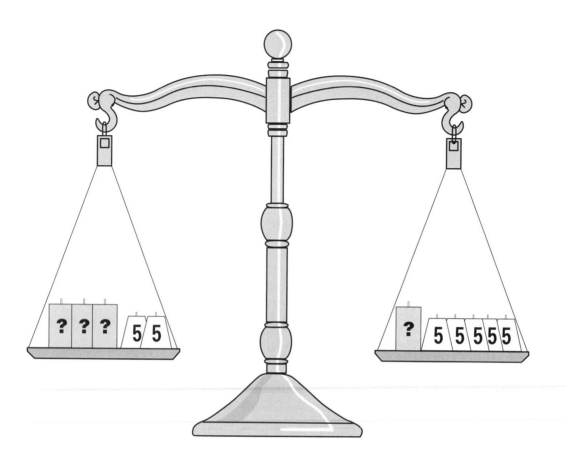

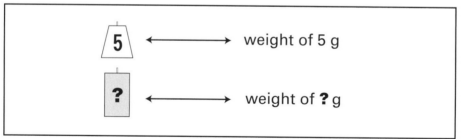

The weights are perfectly in balance.

• Calculate the weight marked by **?**.

On a calendar for June 2004, a horizontal block of three consecutive dates is chosen. The sum of the three dates is 48 (15 + 16 + 17 = 48).

Sun	Mon	Tue	Wed	Thu	Fri	Sat
		1	2	3	4	5
6	7	8	9	10	11	12
13	14	15	16	17	18	19
20	21	22	23	24	25	26
27	28	29	30			

Another set of consecutive dates is chosen. The sum of the dates is 72.
• Which three dates are chosen?

Take another month. On the calendar for that month, three consecutive dates are chosen. Their sum is 39.
• Which dates are chosen?

Go back to the calendar for June 2004. On the calendar, a *vertical* block of three dates is chosen. Their sum is 54.
• Which dates are chosen?

• Is there a vertical block of *four* dates with sum 54? If so, which one?

The third Tuesday in May 2004 was the 15th. This is the smallest possible number that the third Tuesday in any month can have.
• Explain why.

• What is the largest possible number the third Tuesday of a month can have? Show your work.

Calendar Problems (2)

On the calendar of a month with 30 days, a horizontal block of *five* consecutive dates is chosen. The sum of the five numbers is 105.

Sun	Mon	Tue	Wed	Thu	Fri	Sat
			x			

- Use *x* for the middle number and make expressions for the other four numbers.

- Calculate the value of *x*. (Hint: Use an equation.)

- Look at the calendar. Which day is the first of that month?

On the calendar for another month, a *vertical* block of *four* dates is chosen. The sum of the four numbers is 54.

- Use *x* for the smallest number and make expressions for the other three numbers.

- Calculate the value of *x*.

Mathematics in Context

$v + 14 + v = 25 + v + 14 \longrightarrow v = $ _____

$w + 20 + w = 25 + w + 30 \longrightarrow w = $ _____

$x + 28 + x = 14 + x + 14 \longrightarrow x = $ _____

$y + y + 10 + y = 60 + y \longrightarrow y = $ _____

$z + 60 + z + z = 10 + z \longrightarrow z = $ _____

Solving Equations (2)

$$24a + 12 = 20a + 52$$

$\downarrow$

$a = $ _____

$$204a + 121 = 200a + 521$$

$\downarrow$

$a = $ _____

$$2004a + 1218 = 2000a + 5218$$

$\downarrow$

$a = $ _____

$$35b - 11 = 30b - 6$$

$\downarrow$

$b = $ _____

$$350b - 110 = 300b - 60$$

$\downarrow$

$b = $ _____

$$3500b - 1100 = 3000b - 600$$

$\downarrow$

$b = $ _____

On this page, you see eight equations. They can be divided in two groups in such a way that all equations in each group have the same solution. Try to decide if two equations belong to the same group without solving the equations!

- Choose one equation and connect it with all the equations that belong to the same group.
- Check to see whether the remaining equations also belong to the same group.

$$4a + 17 = 3a + 22$$

$$4a + 17 = 3a + 21$$

$$4a + 15 = 3a + 20$$

$$3a + 17 = 2a + 22$$

$$8a + 34 = 6a + 42$$

$$40a + 170 = 30a + 220$$

$$4a + 17 = 3(a + 7)$$

$$2a + 7\tfrac{1}{2} = 1\tfrac{1}{2}a + 10$$

(Dis)covering (1)

Example:

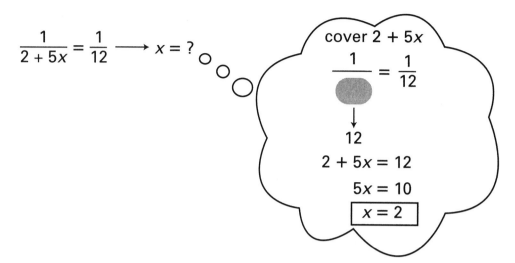

$$\frac{1}{2 + 5x} = \frac{1}{12} \longrightarrow x = ?$$

cover 2 + 5x

$$\frac{1}{} = \frac{1}{12}$$

12

2 + 5x = 12

5x = 10

$$x = 2$$

Use the "cover" method to solve.

$$\frac{1}{8 - x} = \frac{1}{3}$$

$$\frac{1}{3 - x} = \frac{1}{8}$$

$$\frac{5}{\frac{1}{2}x + 1} = \frac{5}{6}$$

$$\frac{6}{2x + 1} = \frac{3}{4}$$

$$\frac{100}{3x + 4} = 10$$

$$\frac{1000}{31x + 69} = 10$$

Solve for **x**.

$$\sqrt{x} = 5 \qquad\qquad 4\sqrt{10 - x} = 12$$

$$\sqrt{4 + 3x} = 5 \qquad\qquad \sqrt{10 + \sqrt{x}} = 4$$

$$2 + \sqrt{3x} = 5 \qquad\qquad \sqrt{30 + \sqrt{30 + x}} = 6$$

$$\sqrt{10 + x} = 3 \qquad\qquad \sqrt{20 + \sqrt{20 + \sqrt{x}}} = 5$$

Name _____ **Date** _____ **Class** _____

Problems of Diophantos (1)

Diophantos was a Greek mathematician who lived in Alexandria in the third century. He wrote a book called *Arithmetica* that has 130 problems about numbers, which are solved by using equations. He may have been the first one who used a symbol for unknown numbers, and, therefore, he is sometimes called the "father of algebra."

Diophantos's epitaph:

> Here lies Diophantos, a math wonder to behold,
> With the use of algebra, the stones tell how old:
> One-sixth of his life, he spent as a child,
> One-twelfth more, and his beard grew wild,
> Adding one-seventh more to his life,
> He decided to marry, and so took a wife,
> Five years passed, and he had a bouncing boy,
> Life seemed perfect, filled with love and joy.
> But alas, the dear son of master and sage,
> After reaching half of his father's full age,
> Died and was laid in a cold, chilly grave,
> While his father, trying his best to be brave,
> Busied himself in the science of numbers,
> Then in four years, enjoyed Eternal Slumbers.

- Try to find out to what age Diophantos lived.
  (Hint: Suppose this age is *a*; use *a* to write an equation.)

Diophantos used a fixed symbol for a variable number
or an unknown number.
This symbol looks like our letter **s**.

Here is problem 3 from the first chapter of his book.
From two numbers we know that their sum is 80.
The largest number is 3 times the smaller one plus 4.
Determine both numbers.

Here is the beginning of Diophantos's solution.
Suppose the smallest number is **s**.
Hence the larger one is $3s + 4$.

- Complete the solution.

The two numbers are _____ and _____ .

Here is problem 5 from Diophantos's book.
From two numbers we know that their sum is 30.
5 times the first number plus 3 times the second num-
ber is equal to 100.

Suppose the first number is **s**. Then the second number
is $30 - s$.

- Complete the solution.

The two numbers are _____ and _____ .

Problems of Diophantos (3)

Here is problem 6 from Diophantos's book.

From two numbers we know that their difference is 20.

6 times the smallest number plus 4 times the larger one is equal to 100.
What are those numbers?

• Complete the solution.

Suppose the smallest number is *s.*
Hence the larger number is _____ .

The two numbers are _____ and _____ .

Here is problem 7 from Diophantos's book.

Subtracting 20 from a required number the result will be 3 times the result of subtracting 100 from the same number.

What is this number?

• Solution

The required number is _____ .

Mathematics in Context

Ronald is saving money for a mountain bike. His parents gave
him $20 to start.

Suppose he saves $15 each month.

• Fill in the table and draw the corresponding graph.

Month	Amount (in dollars)
0	20
1	_____
2	_____
3	_____
4	_____
5	_____

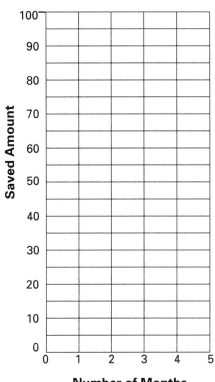

Use **s** for the saved amount and **m** for the number of months.

• Make a formula representing the relationship between
  m and **s**.

• How does the graph change if Ronald saves $10 each month?

• What is the formula if Ronald saves $10 each month?

• How does the graph change if Ronald's parents give him $30
  as a starting amount and Ronald saves $15 each month?

• What is the formula for the case above?

Table, Graph, and Formula (2)

An Internet company offers quick DSL connections that are very useful for e-mailing, surfing, and chatting.

To make the deal more attractive, the company makes an offer:

The first three months are free.

After that you pay $40 each month.

- Complete the table and draw the corresponding graph.

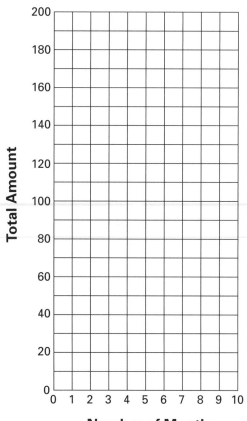

Months	Total Amount Paid (in dollars)
0	0
1	_____
2	_____
3	_____
4	40
5	_____
6	_____
7	_____
8	_____
9	_____
10	_____

- After how many months is the cost $600?

Use *a* for the total amount paid and *m* for the number of months.

- Explain how the formula $a = 40(m - 3)$ represents the relationship between *m* and *a* after three months.

- How does the graph change if the monthly amount is $45 and the first 4 months are free? How does the formula change in this case?

Name _____ Date _____ Class _____

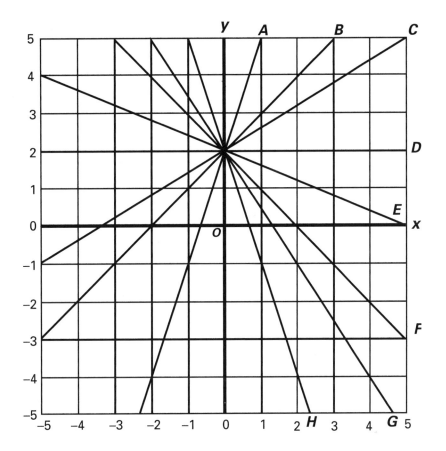

• Give the slope of each of the lines.

line	A	B	C	D	E	F	G	H
slope								

The figure with eight lines above is not symmetric.

• How many lines do you need to add to make the figure symmetric?
  What are the slopes of these lines?

Slopes (2)

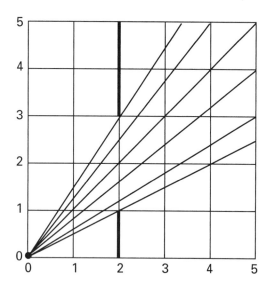

The heavy vertical line in the picture has a gap between the points (2, 1) and (2, 3).

Many rays start at the point (0, 0) and pass through that gap.

• What can you tell about the slopes of all those rays?

If you take another starting point, say (1, 0), many other rays can be drawn that pass through the gap.

• Fill in the table.

Starting Point	Slope Between
(0,1)	_____ and _____
(0,2)	_____ and _____
(0,3)	_____ and _____
(0,4)	_____ and _____
(0,5)	_____ and _____

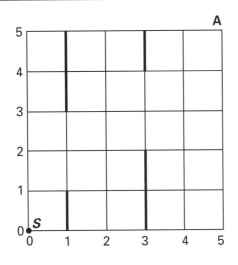

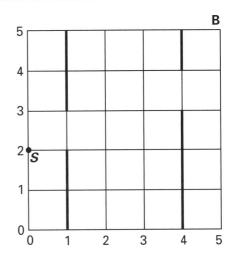

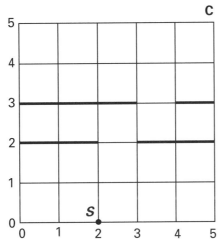

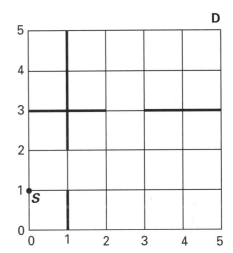

In each of the pictures you see two lines (horizontal or vertical) with a gap.
A pencil of rays starts at **S** and passes through *both* gaps.

• What is the range of the slope? Fill in the table.

Picture	Slope Between
A	_____ and _____
B	_____ and _____
C	_____ and _____
D	_____ and _____

Slopes (4)

From the point (0, 0), a graph is drawn that consists of line segments with slopes $1, \frac{1}{2}, \frac{1}{3}, \frac{1}{4}$.

The graph will be continued in this way up to and including a segment with slope $\frac{1}{10}$.

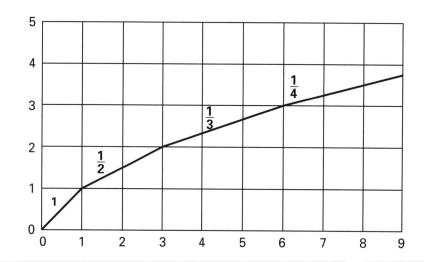

- What are the coordinates of the farthest point?
  Show your work.

Slope and Intercepts (1)

A line with slope $\frac{1}{2}$ passes through point **S** with coordinates (20, 18).

You can think of moving along the line one step at a time.

Each step is a move of −2 units horizontally and −1 unit vertically.

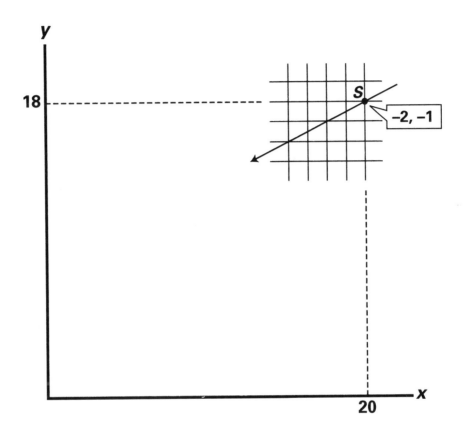

• How many steps from **S** is the y-axis?

• What number is the y-intercept of the line? Explain your answer.

• How many steps from **S** is the x-axis?

• What number is the x-intercept of the line? Explain your answer.

Slope and Intercepts (2)

A line with slope 3 is passing through the point **S**.

A movement along the line goes −1 unit horizontally and −3 units vertically.

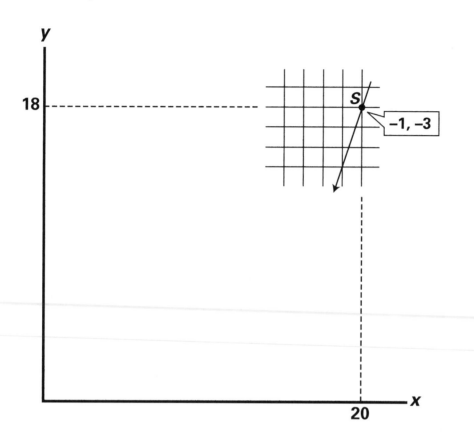

- How many steps from **S** is the x-axis?

- Which number is the x-intercept of the line?
  Explain your answer.

- How many steps from **S** is the y-axis?

- Which number is the y-intercept of the line? Explain
  your answer.

A line is passing through the points **S** and **T** respectively with coordinates (52, 36) and (48, 35).

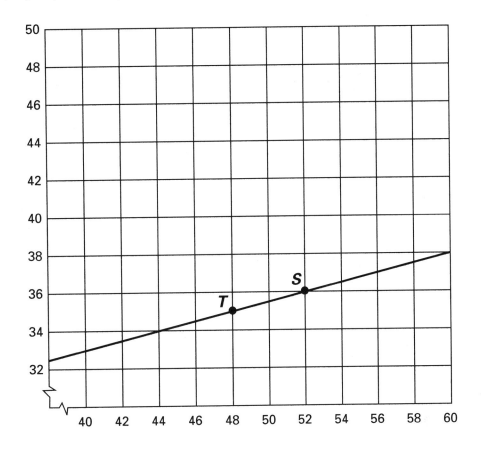

- What is the slope of line **ST**?

- Calculate the *y*-intercept and the *x*-intercept of line **ST**.

Another line passes through the points (44, 36) and (52, 34).
- What is the slope of this line?

- Calculate the *y*-intercept and the *x*-intercept of this line.

Name _____ **Date** _____ **Class** _____

Slope, Intercept, and Equation (1)

Remember:

If you know the *y*-intercept and the slope of a line, you can easily find the equation of that line.

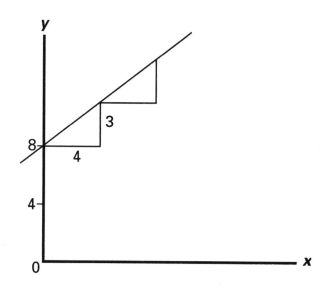

Example: *y*-intercept = 8, slope = $\frac{3}{4}$,   equation: $y = 8 + \frac{3}{4}x$

• Complete the table.

slope	*y*-intercept	*x*-intercept	equation
	24	-8	
$\frac{3}{4}$	9		
$\frac{3}{4}$		−8	
	10	10	
			$y = 10 + 2x$
			$y = 2(x + 5)$
10		1	
			$y = \frac{5}{8}x$

Mathematics in Context

Name _____ Date_____ Class_____

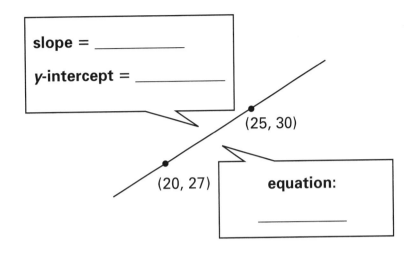

slope = _____

y-intercept = _____

(25, 30)

(20, 27)

equation:

_____

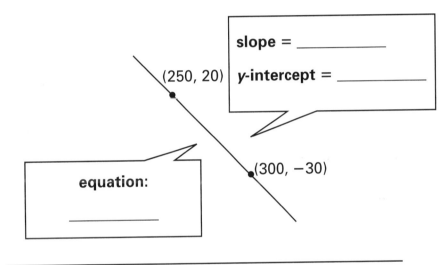

slope = _____

y-intercept = _____

(250, 20)

(300, −30)

equation:

_____

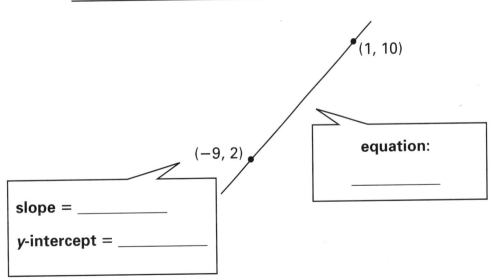

(1, 10)

equation:

_____

(−9, 2)

slope = _____

y-intercept = _____

Name _____ Date_____ Class_____

Intersecting Lines (1)

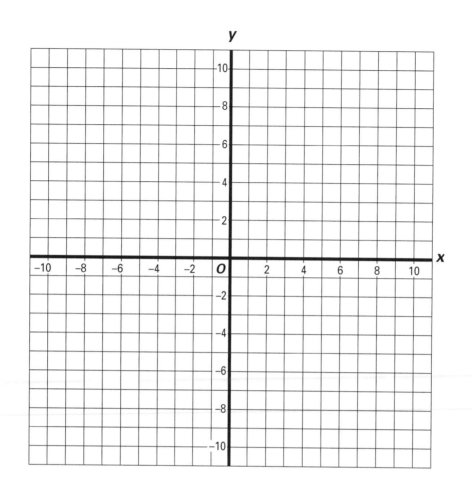

- Draw line **A** that connects the points (−2, 10) and (10, −8).

  What is the equation of that line?

- Draw line **B** that connects the points (10, 9) and (−2, −9).

- What is the equation of that line?

- Calculate the intersection point of **A** and **B**.

Mathematics in Context

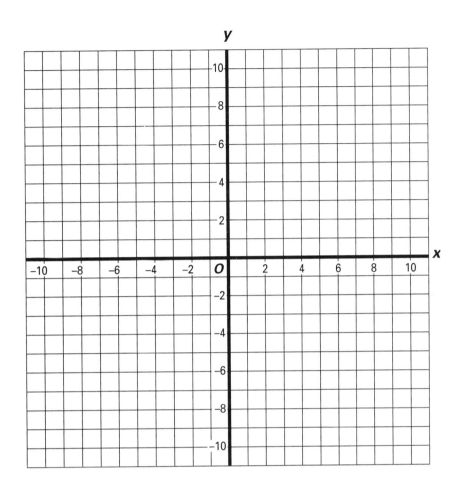

- Draw the line **C** for the equation $y = 3(x - 2\frac{1}{3})$

- What does the number $2\frac{1}{3}$ mean for that line?

- Draw line **D** for the equation $y = -2(x - 3\frac{1}{2})$

- What is the *x*-intercept of that line?

- Calculate the intersection point of **C** and **D**.

Expressions and Weights (1)

$$a + a + b + b + b + c + c + c + c = 2a + 3b + 4c$$

nine terms   three terms

2, 3, and 4 are the **weights** of a, b, and c.

With the weights 2, 3, and 4 and the letters a, b, and c, you can make other expressions, for example: **4a + 3b + 2c**.

There are six different expressions using a, b, and c with weights 2, 3, and 4.

• Write the other four expressions.

• Add the six expressions and you get a new expression with three terms. What is it?

Suppose it is known that **c = b**.

Now you can change **2a + 3b + 4c** into an expression with two terms.

```
┌─────────────────┐
│   2a + 3b + 4c  │
│      c = b      │
└─────────────────┘
        ▼
┌─────────────────┐
│   2a + 3b + 4b  │
└─────────────────┘
        ▼
┌─────────────────┐
│     2a + 7b     │
└─────────────────┘
```

You can do the same with the other five expressions.
• How many different expressions do you get? What are they?

• If you also know that **a = b**, you can simplify these expressions further.

What is the result?

$$w + w + x + y + y + y + y + y + z + z = 2w + x + 5y + 2z$$

w has weight 2, x has weight 1, y has weight 5, and z has weight 2.

The weight 1 is often left out in expressions, but you may want to write **$2w + 1x + 5y + 2z$.**

The sum of the weights in the expression above is 10.

- Give five other expressions in w, x, y, and z, for which the sum of the weights is equal to 10.

- Add those five expressions.
  What is the sum of the weights of the resulting expression?

If it is known that **$w = x$** and **$z = y$**, then **$2w + x + 5y + z$** can be simplified to an expression with two terms:

$$2w + x + 5y + 2z$$
$$w = x \text{ and } z = y$$

$$2x + x + 5y + 2y$$

$$3x + 7y$$

- Simplify your five expressions in a similar way.

The sum of all these five expressions in x and y can also be written as an expression with two terms.

- What is it?

Name ______________________ Date ____________ Class ____________

Powerful Tables (1)

Below you see a table with powers of 2.

n	2^n
0	1
1	2
2	4
3	8
4	16
5	32
6	64
7	128
8	256
9	512
10	1,024
11	2,048
12	4,096
13	8,192
14	16,384
15	32,768
16	65,536
17	131,072
18	262,144
19	524,288
20	1,048,576

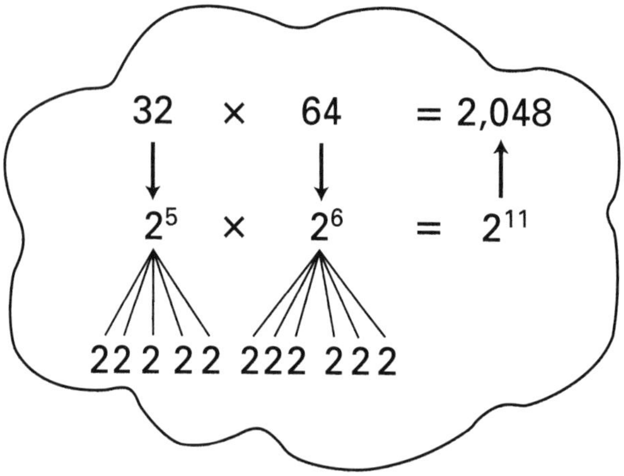
32 × 64 = 2,048
$2^5 \times 2^6 = 2^{11}$
22 2 22 222 222

- Use the table to find the results of the following products.

  16 × 8,192 = ______________

  8 × 16,384 = ______________

  512 × 512 = ______________

  1,024 × 1,024 = ______________

- Use the table to write five different pairs of two positive integers, with a product equal to 1,048,576.

The powers of 3

n	3^n
0	1
1	3
2	9
3	27
4	81
5	243
6	729
7	2187
8	6561
9	19,683
10	59,049
11	177,147
12	531,441
13	1,594,323
14	4,782,969
15	14,348,907
16	43,046,721
17	129,140,163
18	387,420,489
19	1,162,261,467
20	3,486,784,401

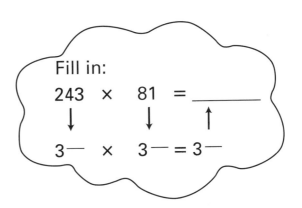

Fill in:

243 × 81 = _____

3— × 3— = 3—

- Find the results of the following products using the table.

81 × 19,683 = _____

2,187 × 59,049 = _____

6,561 × 6,561 = _____

729 × 729 × 729 = _____

- Find the results of the following powers using the table.

$81^3 =$ _____

$243^4 =$ _____

$27^5 =$ _____

- Which number is smaller, 9^{10} or 10^9 ? Explain.

Powerful Tables (3)

- Make a table with powers of 5 up to 5^{10}.

  Write some problems that you can solve using this table.

n	5^n
0	1
1	5
2	
3	
4	
5	
6	
7	
8	
9	
10	

- The table with powers of 1 is very simple. Why?

- Do you know a table of powers with sharp rising results that is very easy to write? Which one?

- If you put the tables with powers of 2 and 3 next to each other, and if you multiply the numbers on the same line, then you get:

$$1 \times 1 = 1$$
$$2 \times 3 = 6$$
$$4 \times 9 = 36$$
$$8 \times 27 = 216$$

$$\dots$$

The results are the powers of 6.

You can check this using your calculator.

- Without a calculator, explain why $2^{10} \times 3^{10} = 6^{10}$.

$$a \times a \times a \times b \times b \times c = a^3 \times b^2 \times c^1 = a^3b^2c$$

The exponents of *a, b,* and *c* are 3, 2, and 1.
(Note: The exponent 1 is usually not written.)

With the exponents 3, 2, and 1 and the letters *a, b,* and *c,*
other products also can be made, for example, ab^3c^2.

Six products can be made using *a, b,* and *c* and the
exponents 1, 2, and 3.

- Write the other four products.

- Multiply the six products together.
  The result can be written in the form $a^{\text{—}} b^{\text{—}} c^{\text{—}}$.
  Which exponents do you get?

If $b = a$, you can simplify a^3b^2c.

```
┌─────────────────┐
│    a³ b² c       │
│    b = a         │
└────────┬────────┘
         ▼
┌─────────────────┐
│    a³ a² c       │
└────────┬────────┘
         ▼
┌─────────────────┐
│    a⁵ c          │
└─────────────────┘
```

You can do the same with the other five products.

- How many different products do you get? Which ones?

- If you also know that $c = a$, you can write each of these
  products as a *power* of *a*. Which one?

Operating with Powers (2)

$$2m^3 \times 3m^2 = 2 \times 3 \times m^5 = 6m^5$$

$2 \times m \times m \times m$ $3 \times m \times m$

• Find as many other multiplications as possible with the same result.

_____ x _____ = $6m^5$

_____ x _____ = $6m^5$

_____

_____

_____

_____

_____

_____

_____

Mathematics in Context

Nicomachos lived in Greece in about the year 100 A.D.

He wrote a book about what he called the "admirable and divine properties" of whole numbers.

Nicomachos sometimes used dots to represent numbers.

Below you see the most famous examples.

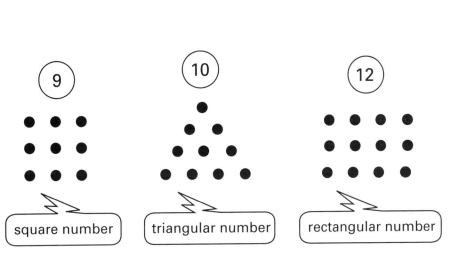
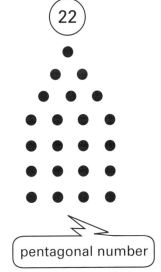

| square number | triangular number | rectangular number | pentagonal number |

He gave every type a geometrical name.

To begin with, consider the family of the **square numbers**.

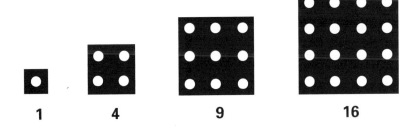

1   4   9   16

- Write the next ten square numbers. You do not need to draw the corresponding patterns, but you can "see" them in your mind.

- Consider the steps between successive square numbers. Do you see any rule? How can you see that rule in the dot pattern?

- 144 is a square number. Is 1,444 a square number? Is 14,444? Use a calculator to investigate this.

Dot Patterns (2)

These are the first four **rectangular numbers**.

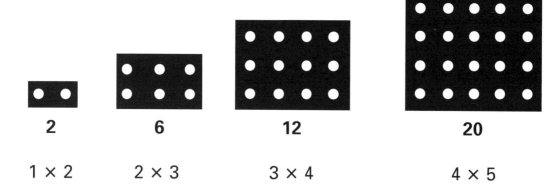

2	6	12	20
1 × 2	2 × 3	3 × 4	4 × 5

- Write the next ten rectangular numbers. Continue the dot pattern in your mind.

- Consider the steps between successive rectangular numbers.
  Do you see any rule? How can you see that rule in the dot patterns?

- Is 9,900 a rectangular number? Explain your answer.

Take the mean of pairs of successive rectangular numbers.

    the mean of 2 and 6 is 4
    the mean of 6 and 12 is 9
    the mean of 12 and 20 is 16

- Continue this at least 5 more times.
  Which special numbers do you get as a result?
  Try to explain your discovery.

Nicomachos was not the first person who used dot patterns for numbers.

Pythagoras lived 600 years earlier (about 500 B.C.). He was a scholar who was the leader of a religious sect.

In the doctrine of Pythagoras, "whole numbers" played a leading part.

He and his disciples had a favorite expression: "Everything is number."

Their favorite number was 10, which is the sum of 1, 2, 3, and 4.

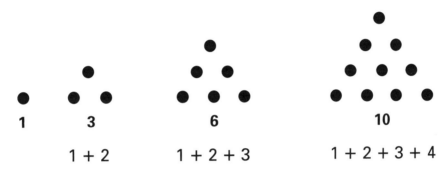

1	3	6	10
	1 + 2	1 + 2 + 3	1 + 2 + 3 + 4

Ten is the fourth number in the sequence of the **triangular numbers**.

• Write the next ten triangular numbers.

The dot patterns of the triangular numbers can also be drawn this way.

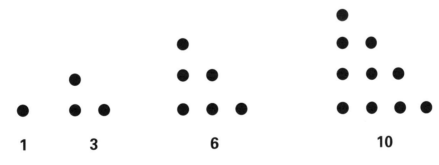

1	3	6	10

• Which special numbers do you get if each of the triangular numbers are doubled? How can you explain this using the dot patterns?

• Is 4,950 a triangular number? Explain your answer.

Strips and Dots (1)

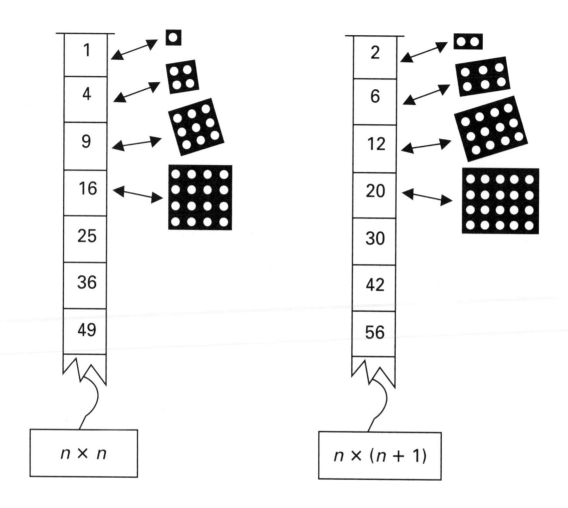

square numbers

rectangular numbers

The expression for the square numbers is usually written as n^2 and read as **n squared.**

- Compare both strips. What strip can you add to the left one to get the right one?

The expressions $n \times (n + 1)$ and $n^2 + n$ are equivalent.

- How can you explain this using dot patterns?

triangular numbers   *rectangular numbers*

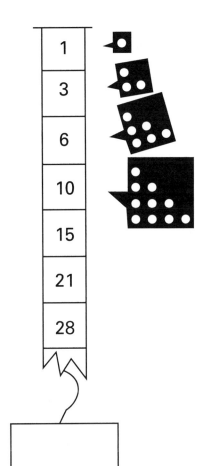

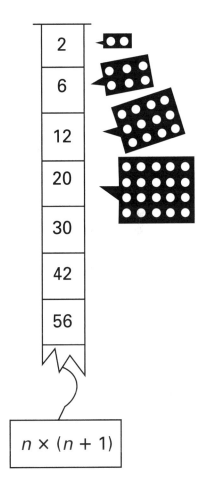

Compare the triangular numbers with the rectangular numbers.

- What expression fits the strip of triangular numbers?

- Give one (or more) equivalent expressions.

Using the expression for triangular numbers, you can calculate the sum of the first hundred positive whole numbers.

$$1 + 2 + 3 + 4 + 5 + \ldots + 98 + 99 + 100 = \underline{\hspace{1cm}}$$

Strips and Dots (3)

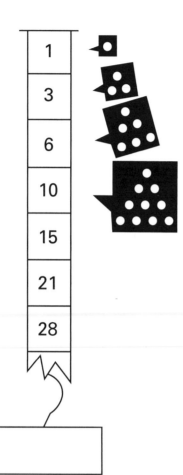

triangular numbers

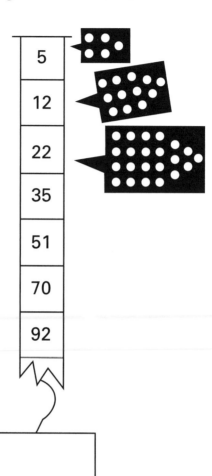

pentagonal numbers

Compare the numbers of both strips.

- Which pentagonal number is right after 92?

- Find an expression that represents the sequence of pentagonal numbers.

Mathematics in Context

Number Spirals (1)

You can make a number line in the shape of a spiral!

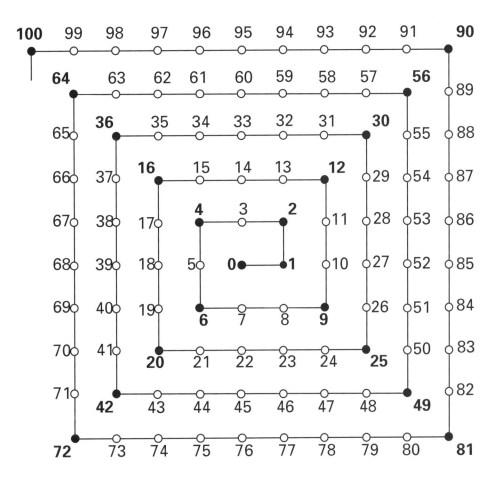

The black vertices of the spiral correspond with special families of numbers.

- Which families?

In the diagram, you can see that every square number is exactly between two rectangular numbers.

For example, **49** is between **42** and **56**.

This is because $49 = 7 \times 7$ and $42 = 6 \times 7$ and $56 = 8 \times 7$.

- The square number 144 lies in the middle between the rectangular numbers _____ and _____ .

- The square number 1,444 lies in the middle between the rectangular numbers _____ and _____ .

- The square number n^2 lies in the middle between the rectangular numbers _____ and _____ .

Number Spirals (2)

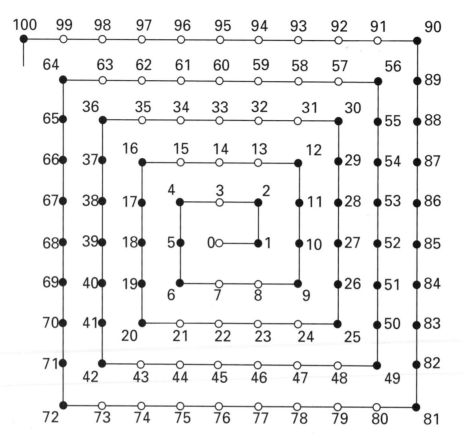

The dots on the vertical parts of the number spiral are black, and the others dots are white. If you add the black numbers from one vertical part, the result is equal to the sum of the next set of larger white numbers.

You can check that.

$$1 + 2 = 3$$
$$4 + 5 + 6 = 7 + 8$$
$$9 + 10 + 11 + 12 = 13 + 14 + 15$$
$$16 + 17 + 18 + 19 + 20 = 21 + 22 + 23 + 24$$

- What is the next line of this pattern?
- You can check this line without calculating both sums.
  (Hint: Mark the steps from "black" to "white."

Line *n* begins with the black number *n*².

- Give an expression for the last black number on that line.

(Hint: How many steps to the end of the black line starting with 25? How many to the end of the black line starting with 64?)

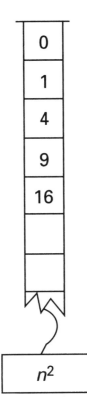

n^2

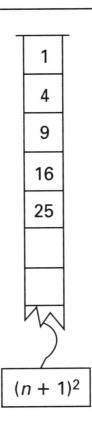

$(n + 1)^2$

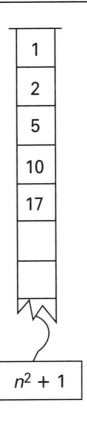

$n^2 + 1$

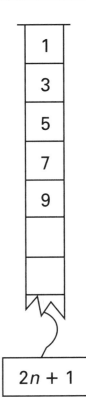

$2n + 1$

• Are these equivalent?

| n^2 | $\overset{?}{=}$ | $n \times n$ |

| $n^2 + 1$ | $\overset{?}{=}$ | $n^2 + 1^2$ |

| $(n + 1)^2$ | $\overset{?}{=}$ | $n^2 + 1^2$ |

| $n^2 + 1$ | $\overset{?}{=}$ | $n \times n + 1 \times 1$ |

| $2n + 1$ | $\overset{?}{=}$ | $n \times n + 1$ |

| $2(n + 1)$ | $\overset{?}{=}$ | $2n + 1$ |

| $(n + 1)^2$ | $\overset{?}{=}$ | $n^2 + 2n + 1$ |

| $(n + 1)^2$ | $\overset{?}{=}$ | $n^2 + 1 + 2n$ |

Strips and Expressions (2)

• Fill in the missing numbers and expressions.

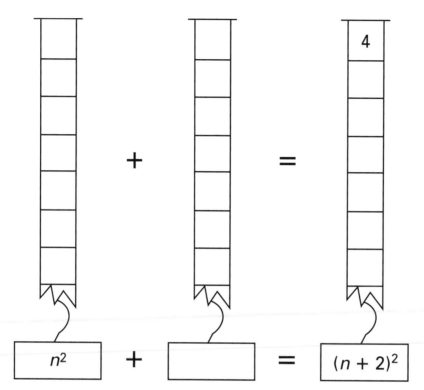

$$n^2 \quad + \quad \boxed{} \quad = \quad (n + 2)^2$$

$$n^2 \quad + \quad \boxed{} \quad = \quad \boxed{}$$

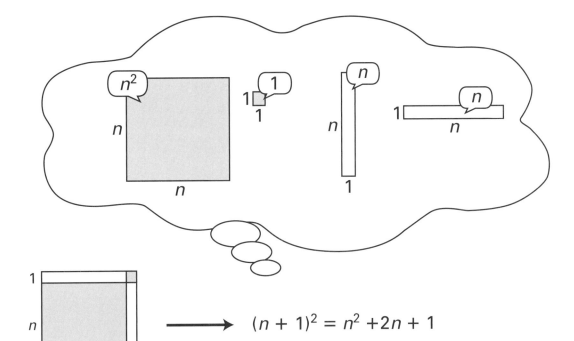

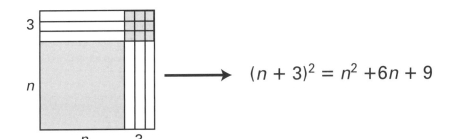

$$(n + 1)^2 = n^2 + 2n + 1$$

$$(n + 2)^2 = n^2 + 4n + 4$$

$$(n + 3)^2 = n^2 + 6n + 9$$

• How will this sequence continue? Write the next five formulas.

Always a Square?

$$1 \times 3 + 1 = 4$$
$$2 \times 4 + 1 = 9$$
$$3 \times 5 + 1 = 16$$
$$4 \times 6 + 1 = 25$$
$$5 \times 7 + 1 = 36$$
$$6 \times 8 + 1 = 49$$

_____

_____

_____

_____

Look at the pattern.

- Write the next four calculations.

The result of each calculation seems to be a square number.

- Try to say what is going on.

- Complete:

29 × 31 + 1 = _____

_____ × _____ + 1 = 2,500

- Write a formula that represents all these calculations:

n × _____ + 1 = _____ .

- How can you explain this formula?

$$32 \times 38 = 30 \times 30 + 10 \times 30 + 2 \times 8 = 1,216$$

$$\underbrace{}_{900} \quad \underbrace{}_{300} \quad \underbrace{}_{16}$$

This shows an example of smart multiplying.

- How can you explain this method by using a 38 × 32 rectangle?

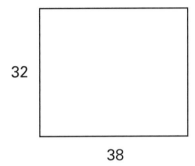

- Calculate in the same way:

  24 × 26 = _____ + _____ + _____ = _____

  43 × 47 = _____ + _____ + _____ = _____

  61 × 69 = _____ + _____ + _____ = _____

  75 × 75 = _____ + _____ + _____ = _____

- Explain at least one of these multiplications by a rectangle.

- Using two numbers between 10 and 100, write three other multiplications you can calculate in the same smart way.

Rectangular Multiplying (2)

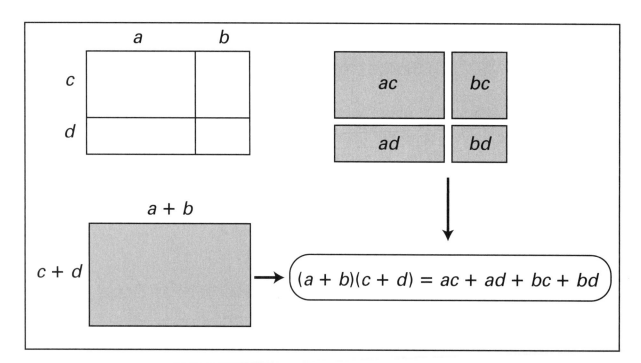

The diagram shows an algebra rule:

The product of the sums **a + b** and **c + d** is equal to the sum of the products **ac, ad, bc,** and **bd**.

- Use this rule to calculate:

  $43 \times 57 = (40 + 3) \times (50 + 7) =$ _____ = _____

  $47 \times 53 =$ _____ = _____ = _____

- Use the same rule to calculate:

  $102 \times 104 =$ _____

  $201 \times 401 =$ _____

  $25 \times 35 + 25 \times 65 + 75 \times 35 + 75 \times 65 = 10,000$

- How can you check this answer without calculating the four products?

- Calculate in a smart way.

  $145 \times 11 + 145 \times 89 + 55 \times 11 + 55 \times 89$

- Explain by using rectangles that the expressions

  (u + v)(x + y + z) and **ux + uy + uz + vx + vy + vz** are equivalent.

- Fill in the blanks for each rectangle.

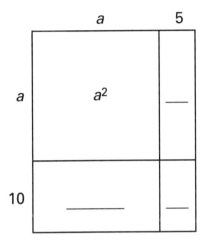

$(a + 5)(c + 10) =$

$ac +$ _____ $+$ _____ $+$ _____

$(a + 5)(a + 10) =$

$a^2 +$ _____ $+$ _____ $+$ _____

The last expression can be simplified to an expression with only three terms.
- How?

The expressions $(x + 3)(x + 7)$ and $x^2 + 10x + 21$ are equivalent.
- Explain why.

- Which pairs of expressions are equivalent? Explain your answers.

  $(a + 4)(b + 6)$ and $ab + 24$

  $(a + 4)(a + 6)$ and $a^2 + 10a + 24$

  $(b + 4)(b + 6)$ and $b(b + 10) + 24$

You Can Count on It (1)

$$1 \times 2 - 0 \times 3 = 2$$
$$2 \times 3 - 1 \times 4 = 2$$
$$3 \times 4 - 2 \times 5 = 2$$
$$4 \times 5 - 3 \times 6 = 2$$
$$5 \times 6 - 4 \times 7 = 2$$

and so on...

- Check the calculations on the blackboard. Continue the sequence with some more lines. What do you think?

- Use numbers between 100 and 1,000 to give some other calculations that fit in the sequence. Check to see if the result is 2.

This is the general rule.

$$(n + 1) \times (n + 2) - n \times (n + 3) = 2$$

- Use the pictures in the cloud to explain that this is true.

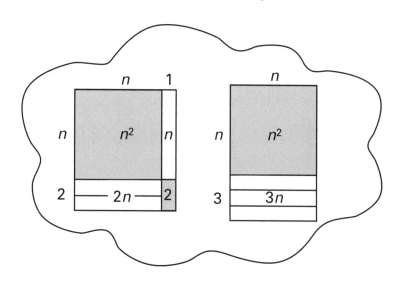

$$1 \times 3 - 0 \times 4 = \underline{\quad}$$
$$2 \times 4 - 1 \times 5 = \underline{\quad}$$
$$3 \times 5 - 2 \times 6 = \underline{\quad}$$
$$4 \times 6 - 3 \times 7 = \underline{\quad}$$
$$5 \times 7 - 4 \times 8 = \underline{\quad}$$

and so on...

- What is the regularity in the sequence of calculations?

- Which formula corresponds to this sequence?

- Draw a picture that explains the formula.

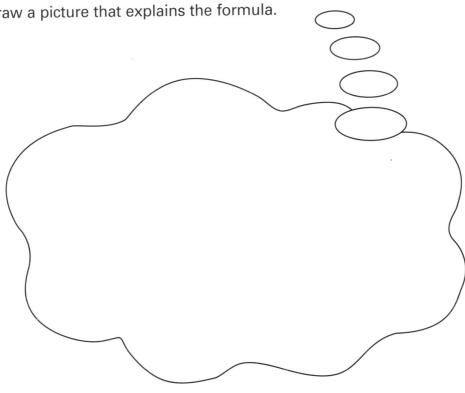

You Can Count on It (3)

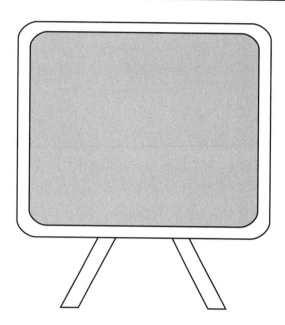

- Design a similar sequence of calculations with the same result on each line.

- Write the formula that corresponds with this sequence.

$$(n + 1) \times (n + 2) = n^2 + 3n + 2$$

$$n \times (n + 3) = n^2 + 3n$$

$(n + 2) \times (n + 3) =$ _____

$n \times (n + 5) =$ _____

— _____

$(n + 2) \times (n + ...) =$ _____

$n \times (n + ...) =$ _____

— 10

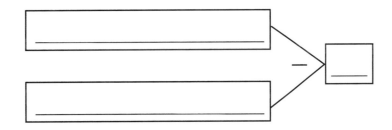

— _____

A Remarkable Identity (1)

A rectangular field has a length of 31 m and a width of 29 m.

- Give a quick estimation (in m²) of the area.
- How many square meters does this estimation deviate from the exact area?
- Answer the same two questions for a field of 41 m by 39 m.

51 × 49 ⟨ estimation: 50 × 50 = _____
          exact result = _____ ⟩ deviation = _____

- Make a similar diagram for 61 × 59.

- Explain using the picture

> The difference between **$n \times n$** and $(n + 1) \times (n - 1)$ is equal to **1**.

or using an equation.

> $(n + 1) \times (n - 1) = n^2 - 1$

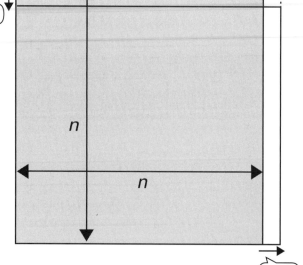

- 32 × 28 { estimation: 30 × 30 = _____
              exact result = _____ } deviation = _____

- Make a similar diagram for 42 × 38.
- Make a diagram for 52 × 48.
- Find a general rule.

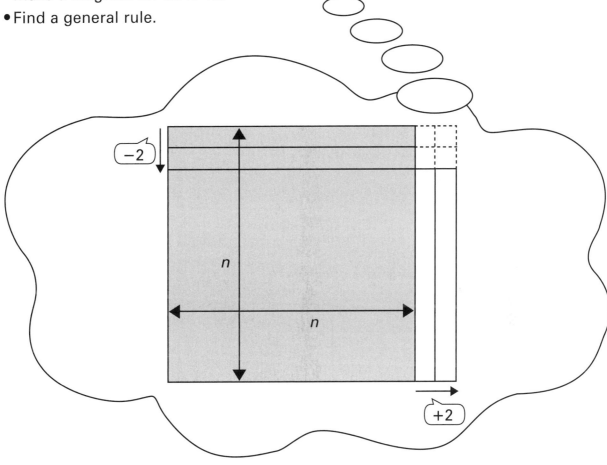

A Remarkable Identity (3)

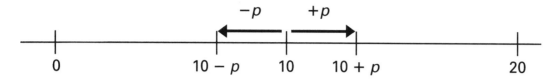

• Use the number line to complete the following.

$(10 + p) + (10 - p) =$ _____

$(10 + p) - (10 - p) =$ _____

August thinks that $10 + p$ times $10 - p$ is equal to 100.

His reasoning: **10** times **10** is **100**
                 10 − **p** is **p** less than **10**,
                 but **10** + **p** is **p** more than **10**,
                 so they compensate each other.

• Is August right? Explain your reasoning.

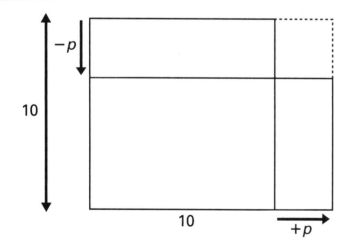

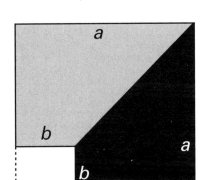

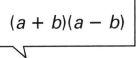

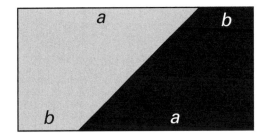

=

$$a^2 - b^2 = (a + b) \times (a - b)$$

Example:

$$54^2 - 46^2 = (54 + 46) \times (54 - 46) = 100 \times 8 = 800$$

- Calculate in the same way as the example. Do not use a calculator.

$52^2 - 48^2 =$ _____ × _____ = _____

$67^2 - 33^2 =$ _____ × _____. = _____

$501^2 - 499^2 =$ _____ × _____ = _____

- Design some more exercises that can be done in the same style.
- **100** times **100** is more than **100** + **p** times **100** − **p**.
  How much more?

- n^2 is more than **n** + **10** times **n** − **10**.
  How much more?

Math Bee

in words	in symbols
The sum of **n** and **m** is reduced by **5**.	$n + m - 5$
The sum of **n** and **8** is multiplied by **n**.	$(n + 8) \times n$
The product of **n** and **m** is reduced by **k**.	
The square of the sum of **n** and **m**.	
The sum of two times **n** and 3 times **k**.	
The sum of **n** and 6 is multiplied by the difference of **n** and **6**.	
The product of the third powers of **n** and **m**.	
The square of 5 times **n**.	
5 times the square of **n**.	

Mathematics in Context

Algebra takes time, and time is money.
Here is a detailed "price list."

	Prices:
operations +, −, ×, ÷	1 point each time
squaring	2 points each time
taking to the 3rd power	3 points each time
taking to the 4th power	4 points each time
etc.	etc.
using variables	1 point each time
parentheses and numbers	free

Example 1: What is the price of **3n + m**?

3	**number**	**free**
n	**variable**	**1 point**
3 × n	**multiplication**	**1 point**
m	**variable**	**1 point**
3 × n + m	**addition**	**1 point**
	total price	**4 points**

Example 2: What is the price of **(3n + m)²**?

3n + m	just calculated	**4 points**
(3n + m)²	squared	**2 points**
	total price	**6 points**

• Find the price of each of the following.

$n^2 + 3n$ _____

$n(n + 3)$ _____

$(n + 1)(n + 3)$ _____

$n^2 + 4n + 3$ _____

The Price of Algebra (2)

$n^2 + 3n$ and $n(n + 3)$ are equivalent expressions.
They do not have the same prices yet! (See page 101.)

• Here are pairs of equivalent expressions.
  For each pair, find out which one is cheaper.

$n + n + n + n$ and $4 \times n$ _____

$n \times n \times n \times n$ and n^4 _____

$n + n + n + n$ and $n + 3n$ _____

$(m + 1)^2$ and $(m + 1)(m + 1)$ _____

$(m + 1)^2$ and $m^2 + 2m + 1$ _____

$a^2 \times a^3$ and a^5 _____

$(a^3)^2$ and a^6 _____

• Compare the prices of a^4b^4 and $(a^2b^2)^2$.

Both expressions are equivalent, but the second one is 2 points cheaper.

• Try to find an expression as cheap as possible, that is equivalent to a^4b^4.

You can rewrite n^{15} in different ways by using equivalent expressions.

Here are some possibilities.

$n \times n \times n \times n \times n \times n \times n \times n \times n \times n \times n \times n \times n \times n \times n$

$(n \times n \times n \times n \times n)^3$

$n^{10} \times n^5$

• Which of them is cheaper than n^{15}?

Try to find an equivalent expression with the lowest price.

• Find the cheapest expression equivalent to x^{24}.

Splitting Fractions (1)

Four thousand years ago, mathematicians in Egypt worked with **unit fractions**, such as:

$$\frac{1}{2}, \frac{1}{3}, \frac{1}{4}, \frac{1}{5}, \text{etc.}$$

All of these fractions have a **numerator** of **1**.

A fraction with a numerator different from **1**, can be split up into unit fractions.

Example 1:

$$\frac{3}{4} = \frac{1}{4} + \frac{1}{4} + \frac{1}{4}$$

Splitting into unit fractions becomes less obvious if you want to use as few unit fractions as possible, in this case:

$$\frac{3}{4} = \frac{2+1}{4} = \frac{1}{2} + \frac{1}{4}$$

Example 2:

$$\frac{19}{24} = \frac{12 + 6 + 1}{24} = \frac{1}{2} + \frac{1}{4} + \frac{1}{24}$$

• Try to split up into unit fractions, using as few fractions as possible.

$$\frac{7}{8} = \frac{\Box + \Box + \Box}{8} = \frac{1}{\Box} + \frac{1}{\Box} + \frac{1}{\Box}$$

$$\frac{5}{6} = \Box = \Box$$

$$\frac{13}{16} = \Box = \Box$$

$$\frac{13}{18} = \Box = \Box$$

$$\frac{99}{100} = \Box = \Box$$

If the numerator and denominator of a fraction are divided (or multiplied) by the same number, the value of the fraction does not change.

Example 1:

$\frac{5}{15} = \frac{1}{3}$ (The numerator and denominator are divided by 3.)

$\frac{b}{5b} = \frac{1}{5}$ (The numerator and denominator are divided by **b**.)

The rule is used to split fractions into unit fractions.

Examples:

$$\frac{n+1}{3n} = \frac{n}{3n} + \frac{1}{3n} = \frac{1}{3} + \frac{1}{3n}$$

$$\frac{a+b}{ab} = \frac{a}{ab} + \frac{b}{ab} = \frac{1}{b} + \frac{1}{a}$$

• Split each fraction into as few unit fractions as possible.

$\frac{p+5}{5p} = \boxed{}$ $\frac{p+3}{3pq} = \boxed{}$

$\frac{x+y}{xy} = \boxed{}$ $\frac{k+m+n}{kmn} = \boxed{}$

• Complete each expression.

$\dfrac{\boxed{} + \boxed{}}{ab} = \dfrac{1}{a} + \dfrac{1}{b}$

$\dfrac{\boxed{} + \boxed{} + \boxed{}}{abc} = \dfrac{1}{a} + \dfrac{1}{b} + \dfrac{1}{c}$

$\dfrac{\boxed{} + \boxed{} + \boxed{} + \boxed{}}{abcd} = \dfrac{1}{a} + \dfrac{1}{b} + \dfrac{1}{c} + \dfrac{1}{d}$

Fractions on Strips (1)

• Find the missing numbers and expressions.

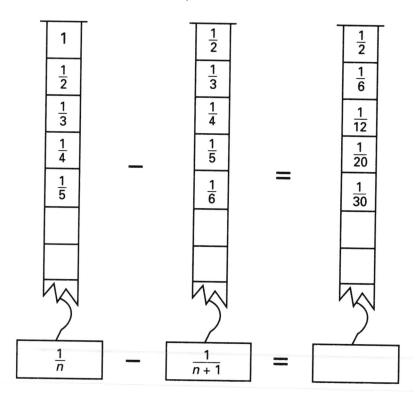

$$\frac{1}{n} - \frac{1}{n+1} = \boxed{}$$

$$\frac{1}{n} + \frac{1}{n+1} = \boxed{}$$

• Find the missing numbers and expressions.

| $\frac{1}{n}$ | $\times$ | $\frac{1}{n+1}$ | $=$ | |

| $\frac{1}{n}$ | $\div$ | $\frac{1}{n+1}$ | $=$ | |

Multiplying Fractions (1)

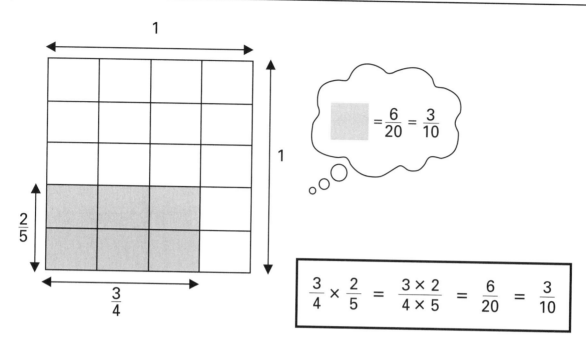

$$\frac{3}{4} \times \frac{2}{5} = \frac{3 \times 2}{4 \times 5} = \frac{6}{20} = \frac{3}{10}$$

- Draw a picture to show that

$$\frac{4}{5} \times \frac{2}{3} = \frac{4 \times 2}{5 \times 3} = \frac{8}{15}.$$

- Calculate

$$\frac{3}{5} \times \frac{1}{4} = \boxed{}$$ $$\frac{3}{5} \times \frac{4}{5} = \boxed{}$$

$$\frac{3}{4} \times \frac{1}{5} = \boxed{}$$ $$\frac{1}{5} \times \frac{1}{4} = \boxed{}$$

- Calculate $p^2 + q^2 + 2pq$ for $p = \frac{1}{3}$ and $q = \frac{2}{3}$.

- Also for $p = \frac{2}{5}$ and $q = \frac{3}{5}$.

- Calculate $p^2 - q^2$ for $p = \frac{3}{4}$ and $q = \frac{1}{4}$.

- Also for $p = \frac{5}{8}$ and $q = \frac{3}{8}$.

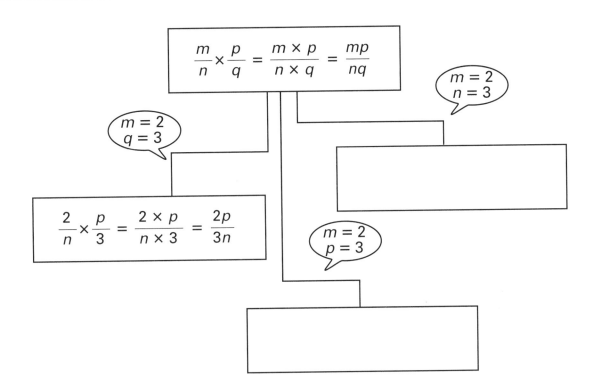

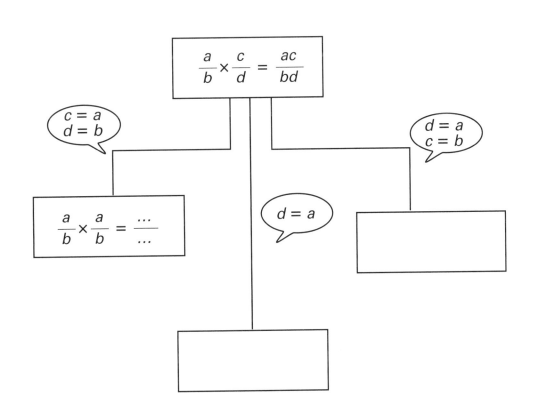

Intermediate Fractions (1)

You have seen that:

$$\frac{m}{n} \times \frac{p}{q} = \frac{m \times p}{n \times q} \qquad \text{(A)}.$$

Maybe you think that addition follows a similar rule.

$$\frac{m}{n} + \frac{p}{q} \overset{?}{=} \frac{m+p}{n+q} \qquad \text{(B)}.$$

- In the same way, calculate the "sum" of $\frac{3}{5}$ and $\frac{1}{4}$.

Check to see whether the result is somewhere *between* $\frac{3}{5}$ and $\frac{1}{4}$.

So it cannot be the real sum of both fractions!

- How can you calculate the sum of $\frac{3}{5}$ and $\frac{1}{4}$?

Formula (B) is not a good recipe to add fractions, but it is a recipe to find intermediate fractions.

- Check for some examples to see if the value of $\frac{m}{n} + \frac{p}{q}$ always lies between the values of $\frac{m}{n}$ and $\frac{p}{q}$.

- Design some examples, for which the value of $\frac{m}{n} + \frac{p}{q}$ is in the middle between the values of $\frac{m}{n}$ and $\frac{p}{q}$.

In a group of 31 students, there are more girls than boys (19 versus 12). In a parallel group (29 students), it is just the reverse (12 versus 17). During physical education, girls and boys of both groups are in separate classes, one all girls class (19 + 12) and one all boys class (12 + 17).

- Check to see if this results in a more balanced distribution of girls and boys.

- What does this example have to do with the concept of intermediate fractions?

Mathematics in Context

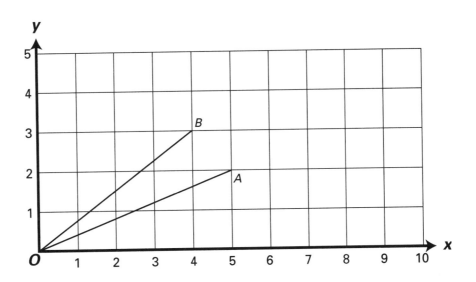

•What fraction represents the slope of the line *OA*?

•What fraction represents the slope of the line *OB*?

•Add the numerators and denominators of both fractions.
 Which fraction do you get?

•Draw a line starting at *O*, with a slope equal to this fraction.
 How can you see whether this fraction is an intermediate
 fraction of the fractions corresponding to the slopes of
 OA and *OB*?

•Do the same with two other lines starting at *O*.
 Determine the corresponding fractions, add the numerators
 and denominators to get a new fraction, and draw the line
 whose slope is equal to the new fraction.

Intermediate Fractions (3)

New fractions can be made step by step using the recipe
to make intermediate fractions.

- Fill the empty cells in the "tree" below.

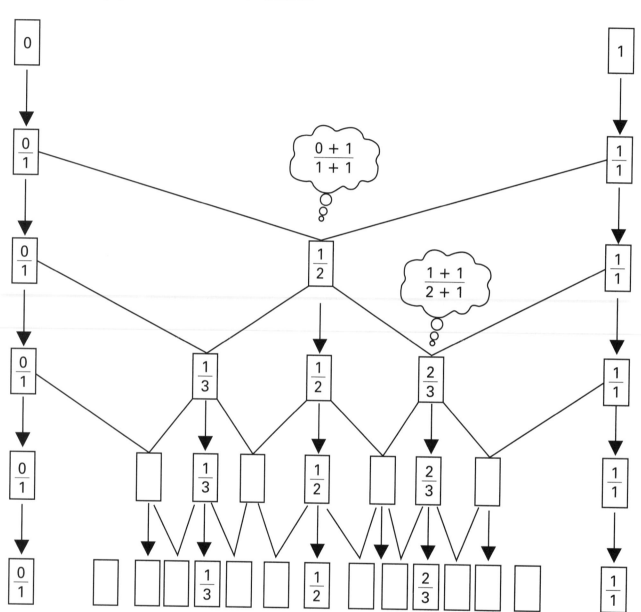

Adding Fractions

The correct formula for adding fractions is (unfortunately) much more complicated than the rule for making intermediate fractions.

Here it is. $\dfrac{m}{n} + \dfrac{p}{q} = \dfrac{m \times q + n \times p}{n \times q}$

- Check this formula for the fractions $\dfrac{3}{5}$ and $\dfrac{1}{4}$.

- Design some more examples to check the addition formula.

- What (simpler) formula do you get if **m = 1** and **p = 1**?

- What (obvious) formula do you get if **n = 1** and **q = 1**?

The addition formula is more complicated than necessary in many cases, but it's always correct.

Here is an example, in which you can add fractions without any formula.

$$\dfrac{3}{n} + \dfrac{2}{n} = \dfrac{5}{n}$$

- If you use the addition formula (with **m = 3**, **p = 2**, and **q = n**), you would find $\dfrac{3}{n} + \dfrac{2}{n} = \dfrac{3 \times n + n \times 2}{n \times n}$.

  Show that the result is equal to $\dfrac{5}{n}$.

- Find one fraction equal to $\dfrac{m}{3} + \dfrac{m}{2}$.

- Do the same thing for $\dfrac{m}{3} + \dfrac{p}{2}$.

Trees with Fractions (1)

• Fill in the missing expressions.

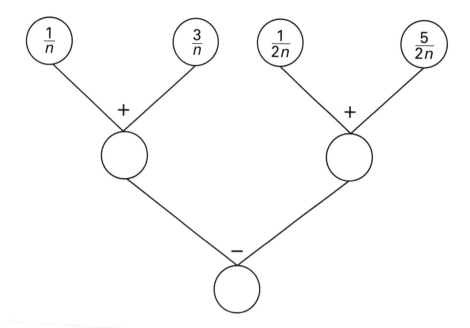

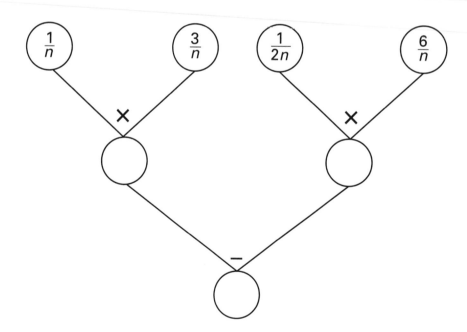

Mathematics in Context

● Fill in the missing expressions.

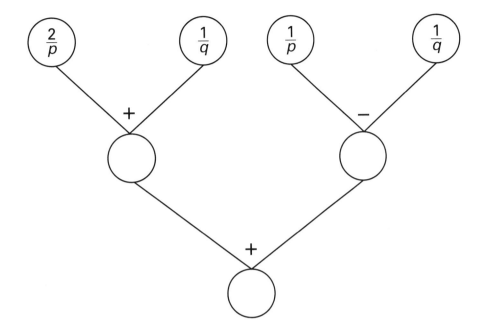

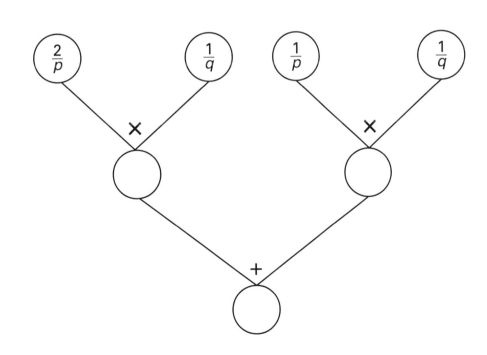

More Fractions

$\dfrac{a}{2} + \dfrac{a}{3} + \dfrac{a}{6} =$ [ ]

$\dfrac{2}{a} + \dfrac{3}{a} + \dfrac{6}{a} =$ [ ]

$\dfrac{b}{2} + \dfrac{b}{4} + \dfrac{b}{6} + \dfrac{b}{12} =$ [ ]

$\dfrac{2}{b} + \dfrac{4}{b} + \dfrac{6}{b} + \dfrac{12}{b} =$ [ ]

$\dfrac{c}{2} + \dfrac{c}{6} + \dfrac{c}{10} + \dfrac{c}{12} + \dfrac{c}{15} + \dfrac{c}{60}$

$\dfrac{2}{c} + \dfrac{6}{c} + \dfrac{10}{c} + \dfrac{12}{c} + \dfrac{15}{c} + \dfrac{60}{c}$

• Design one pair of additions in the same style.

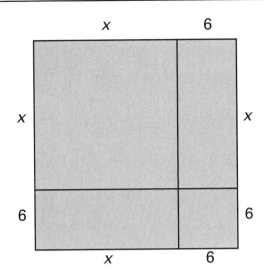

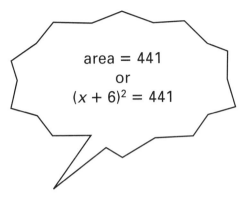

area = 441
or
$(x + 6)^2 = 441$

• Calculate the value of x.

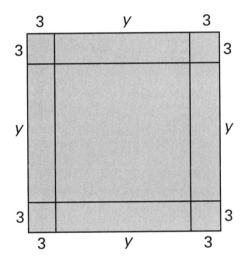

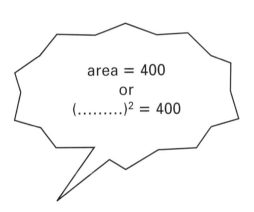

area = 400
or
$(.........)^2 = 400$

• Calculate the value of y.

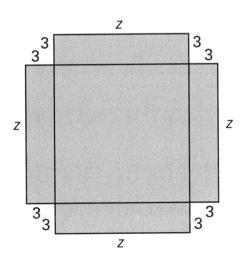

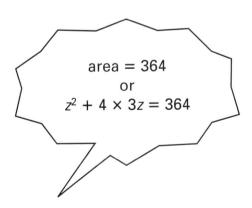

area = 364
or
$z^2 + 4 \times 3z = 364$

• Calculate the value of z.

Equations with Squares (2)

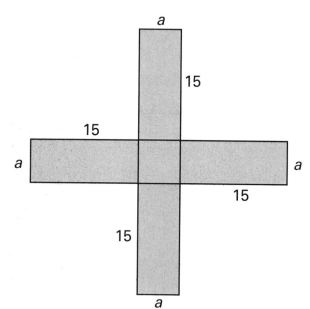

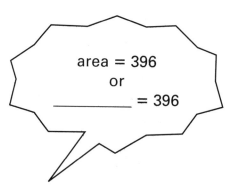

area = 396
or
_____ = 396

- Calculate the value of *a*.

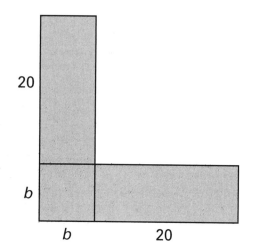

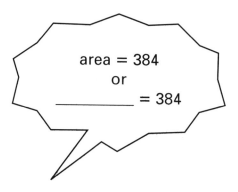

area = 384
or
_____ = 384

- Calculate the value of *b*.

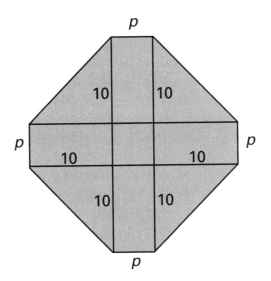

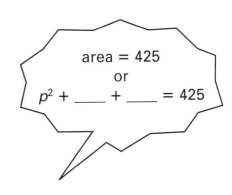

area = 425
or
p^2 + ____ + ____ = 425

• Calculate the value of **p**.

area = 343
or
____ + ____ = 343

• Calculate the value of *x*.

Babylonian Algebra (1)

About 1700 B.C., the Babylonians could solve many algebraic problems.

One type of problem was to calculate two numbers given the sum and the product of those numbers.

Here is an example:

- Find two numbers such that their sum is 18, and the product is 65.

  Solve this problem by trial and error.

You can probably solve this problem rather quickly.

- Find two numbers such that their sum is 180, and the product is 6,500.

- Find two numbers such that their sum is 180, and the product is 7,700.

The next problem is more difficult. (Use your calculator.)

- Find two numbers such that their sum is 180, and the product is 7,776.

Mathematics in Context

The Babylonians didn't use a method of trial and error but discovered an algebraic solution.

They reasoned more or less like this.

Suppose the sum of two numbers is 180, and the product is 7,776.

If the numbers were equal, each of them should be 90.

But $90 \times 90 = 8,100$, and this is larger than 7,776.

So one of the numbers must be larger than 90, say $90 + d$,

and the other one must be **equally** smaller, therefore $90 - d$.

The product of $90 + d$ and $90 - d$ is equal to $8,100 - d^2$.

• Explain this from the diagram below.

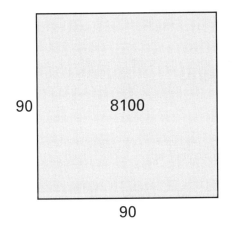
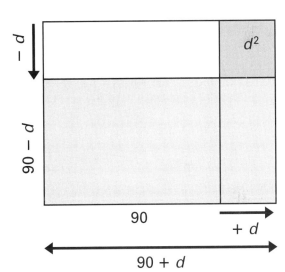

The Babylonians continued.

On the other hand we know that the product must be 7,776, so

$$8100 - d^2 = 7,776$$

• Now calculate the value of **d**.

• What are the two numbers?

Babylonian Algebra (3)

If two numbers are added, the result is 150, and if the same numbers are multiplied, the result will be 5,576.

- Find the two numbers using the Babylonian method.

- Find two numbers if you know that the sum is 7 and the product is $8\frac{1}{4}$.

The perimeter of a certain rectangle is 46 m, and the area is 126 m^2.

- Find the length and the width of that rectangle.

Mathematics in Context

Babylonian Algebra (4)

Another Babylonian problem was to calculate two numbers given the difference and the product of those numbers.

Example:

The difference of two numbers is 12, and the product is 189.

Trial and Error	*Babylonian Method*

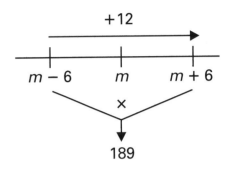

$a(a+12) = 189$

$(m+6)(m-6) = 189$

- Try some values for *a*, until you find one that works.

- Solve this equation and find both numbers.

Patterns and Formulas (1)

A farmer plants apple trees in a square pattern. In order to protect the trees against the wind, he plants conifers all around the orchard.

Here is a diagram of this situation where you can see the pattern of apple trees and conifers for any number (*n*) of rows of apple trees.

```
× × ×            × × × × ×        × × × × × × ×       × × × × × × × × ×
× ● ×            × ●   ● ×        × ●   ●   ● ×       × ●   ●   ●   ● ×
× × ×            ×       ×        ×           ×       ×             ×
                 × ●   ● ×        × ●   ●   ● ×       × ●   ●   ●   ● ×
  n = 1          × × × × ×        ×           ×       ×             ×
                                  × ●   ●   ● ×       × ●   ●   ●   ● ×
                   n = 2          × × × × × × ×       ×             ×
× = conifer                                          × ●   ●   ●   ● ×
● = apple tree                      n = 3            × × × × × × × × ×

                                                           n = 4
```

- How many apple trees does the pattern have when **n = 5**?

  How many conifers does that pattern have?

- Answer the same questions for **n = 10**.

You can use two formulas to calculate the number of apple trees (**A**) and the number of conifers (**C**).
- Complete these formulas.

  A = _____ and **C** = _____

There is a value for **n** for which the number of apple trees equals the number of conifers.
- Find the value of **n** and show your calculations.

There is a value for **n** for which the number of apple trees and conifers together equals 240.
- Find the value of **n**. Show your work.

© Encyclopædia Britannica, Inc. This page may be reproduced for classroom use.

Artist Ivo ten Hove won a contest to design a new tile to be used for sidewalks or patios.

He made a very large X-shaped tile measuring 90 × 90 cm. Here you see a model of this tile.

You can make different square patios with the big tiles and fill the holes with red rectangular tiles as you see in the diagram.

n = 1

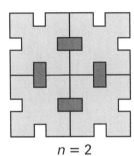

n = 2

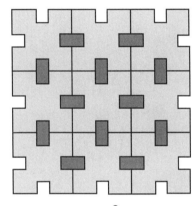

n = 3

- How many big X-shaped tiles are needed for the pattern with **n = 4**?

  How many red tiles?

- How many big tiles are needed for the pattern with **n = 10**?

  How many red tiles?

B represents the number of big X-shaped tiles, and **R** represents the number of red tiles in a square patio.

- Write an expression with **n** that you can use to compute the number of big tiles.

Helen thinks about a formula to calculate the red tiles.

She thinks this is a good one: $R = 2n^2 - 2n$

- Is she right? How did she reason?

Peter has found another formula: $R = 2n(n - 1)$

- Is his formula correct? How did he reason?

Patterns and Formulas (3)

In a factory, tubes are collected in bundles and bound with steel wire.
The diagram shows the way this is done. Front views are shown.

n = 1

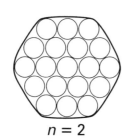
n = 2

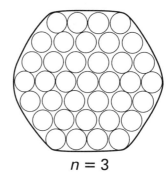
n = 3

There is a smart way to count the tubes in each bundle.
Look at the bundle **n = 3**.

- Fill in the table.

Bundle **n** =	Number of Tubes in Bundle (**T**)
1	7
2	
3	
4	
5	

Rectangular numbers can be used to calculate the number of tubes in a bundle.

- Explain how.

T is the number of tubes in one collection. **T** can be expressed in **n**.

- Write a formula you can use to compute the number of tubes in any bundle.